12 FEB 1977
16 FEB 1980
12 JUL 1980
17 NOV 1980
27 DEC 1980
25 DEC 1982
29 JAN 1983
26 SEP 1983
L 33

D1348937

World Timbers

VOLUME ONE

WORLD TIMBERS

VOLUME ONE

EUROPE & AFRICA

Compiled and edited by

B. J. RENDLE

LONDON: ERNEST BENN LIMITED

UNIVERSITY OF TORONTO PRESS

First published in this form 1969
by Ernest Benn Limited
Bouverie House, Fleet Street, London, EC4
Published in Canada by
University of Toronto Press, Toronto 5, Ontario
8020 1570 0

Printed in Great Britain
510–48001–2

0800 600 900
Bouverie Publishing
0800 671 444
Co.
yellow pages
131 temple chambers
EC4
0171 583 3030
0171 334 7334
p150

Contents

[*continued on page 6*]

Contents [*continued from page 5*]

Introduction

A feature of the journal *Wood*, since it commenced publication in 1936, has been the two series of colour plates of timbers accompanied by technical information on their properties and uses.
The first series, under the name WOOD SPECIMENS, covered the years 1936–1960; this was followed by the current series of WORLD TIMBERS.
A selection of one hundred of the WOOD SPECIMENS was reproduced in book form in 1949[1] and a second volume appeared in 1957[2]. A disadvantage of these volumes, which are now out of print, was the haphazard arrangement of the timbers. The present proprietors of *Wood* have decided to re-issue selected plates from both series (WOOD SPECIMENS and WORLD TIMBERS) in a more systematic arrangement with the technical information revised to bring it up to date and in line with the series now appearing in *Wood*.
This first volume of WORLD TIMBERS covers the continents of Europe and Africa. Volume 2 will deal with North and South America, Central America and the West Indies and Volume 3 with Asia, Australia and New Zealand.
The timbers have been selected primarily for their economic importance or interest on the world market. Many of the species included in the earlier publication are no longer in general use. On the other hand the opening up of new areas of supply, combined with improvements in methods of production and processing and changes in utilisation, have resulted in other species assuming commercial importance.

The Colour Plates

The colour plates have been prepared photographically from specially selected specimens of timber. Care has been taken to ensure that the specimens are representative of the timbers concerned. However, anyone with a knowledge of wood is well aware of the limitations of a single specimen or a single illustration. Allowance must be made for the normal variations in colour, grain and texture, which are so characteristic of wood, as also the differences in appearance due to the method of converting a log (by sawing through-and-through, or by quartering, rotary peeling, slicing, etc.) and the different types of decorative figure that are found in abnormal logs. Some indication of the variation that is to be expected and the changes in colour that occur with the passage of time is given in the description accompanying each plate. It is advisable, however, when choosing wood for high-class decorative work—whether solid timber or veneer—to examine and select material from the merchant's actual stocks. Small samples may be misleading.

The Technical Descriptions

The technical information accompanying the plates is presented in a form designed to appeal to those who are interested in timber but are not specialists in wood technology. The publishers believe that it will be particularly useful to architects and their clients and those members of the timber trade, the timber-using industries and the general public who do not have access to a representative collection of timber specimens and a library of standard reference books on timber.

[1] Wood Specimens—100 Reproductions in Colour. The Nema Press, London, 1949.
[2] A Second Collection of Wood Specimens—100 Reproductions in Colour. The Tothill Press Ltd., London, 1957.

Each timber is described in such detail as its importance seems to warrant. The aim is to indicate its outstanding characteristics and to bring out the practical significance of the information presented.

In selecting a timber for a particular purpose some knowledge of the supply position is essential. The section headed *Distribution and supplies* gives the geographical distribution in broad terms, emphasising the principal sources of supply, with some indication of the quantity and sizes available and whether the timber is commonly stocked in the form of logs, lumber, veneer, etc., or in special sizes such as flooring strips, or is likely to be obtainable only to special order. The notes on supplies refer particularly to Britain. Although the supply position is subject to change and varies in different parts of the world, it is believed this information will be useful to intending purchasers. Before making a final selection, however, it may be advisable to consult a timber merchant.
The main part of each description consists of practical information on the technical properties of the timber in simple language with the minimum of technical terms. The information is largely based on the results of scientific tests carried out by research laboratories and the source of the data is hereby acknowledged. A list of the principal publications consulted—and recommended to those readers who require more detailed information—is on page 187.
The terms used to classify the timbers in respect of certain technical properties are the standard terms adopted by the Forest Products Research Laboratory of Great Britain and defined in the Laboratory's Handbook of Hardwoods (1956) and Handbook of Softwoods (1960).

Seasoning and movement. In a book of this kind it is considered sufficient to state briefly the rate at which a timber dries and the degree of deterioration due to distortion, splitting, etc., to be expected. In some cases the recommended kiln-drying schedule is mentioned; particulars of these schedules are given in Forest Products Research Laboratory Leaflet No. 42. As applied to timber, the term movement means the tendency to shrink or swell in service under varying atmospheric conditions. On the basis of a standard method of measuring the dimensional changes of small timber samples over a predetermined range of atmospheric humidity, timbers are arbitrarily classified as having small, medium or large movement values (for particulars see FPRL Leaflet No. 47).

Strength and bending properties. It is not easy to generalise about the strength of a timber because this term covers a number of specific strength properties, or mechanical properties as they are sometimes called. It has been found more convenient, therefore, to indicate the strength of each timber by comparing it with a well-known standard timber and by mentioning any outstanding mechanical property. Classification according to steam-bending properties is based on the minimum bending radius of sound, clear specimens one-inch thick. Timbers are classified in one of five groups ranging from Very poor to Very good (for particulars see FPRL Leaflet No. 45).

Durability and preservative treatment. The term durability is used to describe the natural resistance of a timber to fungal decay and insect attack. For convenience timbers are roughly classified in five grades intended to indicate the useful life of the timber in contact with the ground, as follows:

Perishable (5 years or less) e.g. ash, beech and the *sapwood* of most timbers.
Non-durable (5–10 years) e.g. elm, obeche, Scots pine, spruce.
Moderately durable (10–15 years) e.g. walnut, African mahogany, Douglas fir, larch.
Durable (15–25 years) e.g. oak, agba, western red cedar, yew.
Very durable (more than 25 years) e.g. teak, iroko.

Resistance to the attack of specific insects (and also marine borers) is mentioned where appropriate.
It is sometimes more convenient to use a non-durable timber treated with a preservative than a naturally durable timber. Some timbers are readily impregnated

with preservatives whereas others are more or less impermeable and cannot be given a satisfactory treatment. Where a long service life is required under conditions favourable to decay or insect attack it is important to choose a timber that will absorb an adequate amount of preservative. The terms used to describe amenability to preservative treatment are self-explanatory. It should be noted that they refer to *heartwood*; the sapwood is usually much more permeable.

Uses. The section on uses cites some of the more typical uses of the timber in question and, where appropriate, indicates its suitability for various purposes in comparison with other species, bearing in mind that suitability may depend as much on economic factors, such as availability, price and sizes, as the technical and aesthetic properties of the timber. This section reflects the editor's personal knowledge of timber utilisation in Britain but much of the information will be found to apply in other countries as well.

Metrication

Great Britain is now committed to the adoption of the metric system in commerce and industry. It will be some time before the change is complete in all fields and during the transitional period both foot/inch measure and metric will be in use for timber. To meet this situation numerical data are expressed in metric units as well as in the traditional form. The figures given are approximate; for greater accuracy the conventional British units of measurement used for timber may be converted to metric units by using the appropriate conversion factor as follows:

To convert inches to mm. multiply by 25·4
" " feet to m. multiply by 0·305
" " lb. to kg. multiply by 0·454
" " lb./ft^3 to kg/m^3 multiply by 16·02

Acknowledgement

Plates on page 15 are Crown copyright. They are reproduced from Forest Products Research Laboratory Leaflet No. 37, *Selecting Ash by Inspection*, by permission of the Controller of H.M. Stationery Office.

European Timbers
HARDWOODS

Alder

[principally *Alnus glutinosa*]

Distribution and supplies. The common alder (*Alnus glutinosa*) is widely distributed in the British Isles and on the Continent. It grows naturally in low-lying marshy places and on the banks of streams and rivers but is not often planted for timber production. In the UK it is usually a small tree, 9–12 m. (30–40 ft.) high and 0·3–0·6 m. (1–2 ft.) in diameter, with a clean bole of 6 m. (20 ft.) or so. The allied species, grey alder (*A. incana*), is common on the Continent. Supplies of alder are limited and mainly in the form of poles up to 250 mm. (10 in.) diameter. Alder is also used for plywood.

General description. The wood resembles birch in colour and texture but is softer, and lighter in weight (density about 0·53 (33 lb./ft.^{3}) seasoned). The cut surface is rather dull and lustreless; it is characterised by lines and streaks like pencil marks, due to the large rays, and scattered rust-coloured flecks (pith flecks). It is pale when first cut, darkening to light reddish-brown; normally there is no visible distinction between sapwood and heartwood. A peculiar feature is that the ends of freshly felled logs assume a characteristic orange-brown colour on exposure to the air. The grain is inclined to be irregular.

Technical properties. For a hardwood, alder is rather soft and weak—comparable to poplar in most of its strength properties—and is not very suitable for bent work. It seasons well, is not resistant to decay but can be easily treated with preservatives. It works well in all hand and machine operations but cross-grained material tends to tear in planing; sharp, thin-edged tools are essential to obtain a smooth finish. It is a fairly good turnery wood, takes nails well and can be glued, stained and painted satisfactorily.

Uses. The timber is most familiar in the form of utility plywood—manufactured largely in Czechoslovakia and the USSR—for making boxes, crates, cheap furniture, etc. In the solid form, being fairly soft and easily shaped, it is the traditional material for making clog soles; it is also considered one of the more suitable timbers for hat blocks, brush backs, and general turnery, including rollers for the textile industry.

Alder

Flat-cut

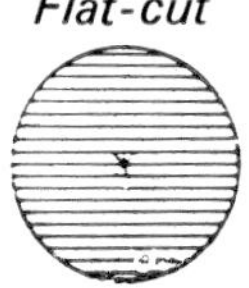

Reproduced two-thirds actual size

Ash

[*Fraxinus excelsior*]

Distribution and supplies. Ash is one of the most valuable European hardwoods; its importance is due to its toughness or resistance to shock, and its excellent bending properties. It is widely distributed on the Continent, and the timber is exported to the UK from France and other European countries to supplement domestic supplies. In favourable situations it grows to well over 30 m. (100 ft.), with a clear bole of 9–12 m. (30–40 ft.) and a diameter of 1 m. (3 ft.) or more. However, the best sports-quality ash is generally obtained from smaller trees. In some parts of Britain coppice-grown ash is used in handles for brooms, garden tools, axes, etc., and for thatching spits.
To meet its extensive range of requirements ash is commonly stocked in a wide range of thicknesses, 25–100 mm. (1–4 in.) and logs are generally available for conversion to special sizes and for veneers. Some timber merchants offer prime square-edged or unedged boards selected for colour.

General description. The wood is typically white to light-brown but temporarily pink when freshly cut. Sapwood and heartwood are not usually well defined, though a dark-brown or black heart is often found in old trees. This is not necessarily a sign of decay but reduces the market value of the timber. The trade name olive ash refers to a variegated form of brown-hearted ash used for decorative veneer and fancy goods.
Ash is a typical ring-porous hardwood, that is to say, the pores of the spring wood are much larger than those of the summer wood, and form a well-defined zone or ring. The summer wood consists mainly of thick-walled strengthening fibres, so the width of this zone is an indication of the strength of the wood. The conspicuous growth rings produce a characteristic decorative figure in flat-sawn timber and rotary-cut veneer. The grain is usually straight; wavy grain (ram's horn or 'rammy' grain) is not uncommon in the butts of old trees.
Depending on conditions of growth, ash is subject to considerable variation in density, from 0·53–0·83 (33–52 lb./ft.^{3}) in the dry condition, average about 0·69 (43 lb.).

Seasoning and movement. Air seasoning presents no difficulty and is fairly rapid, but existing shakes will probably open more. Ash can be dried fairly easily in a kiln if temperatures are kept low, otherwise distortion may occur, sometimes accompanied by end splitting. If distortion develops during kilning the timber will respond to reconditioning with high-temperature steam. Ash is only moderately stable under varying conditions of humidity.

Strength and bending properties. The outstanding property of ash is its toughness or resistance to shock; this is generally highest in wood laid down in the early life of healthy, fast-growing trees with wide rings and a large proportion of dense summer wood. As regards its other strength properties, ash is harder and stiffer than oak and more resistant to splitting, and about equal to oak in bending strength and crushing strength along the grain.
Ash is well known for its excellent steam-bending properties; these are, however, adversely affected by irregular grain and relatively small pin knots. Tests have shown that sound black-heart ash bends as well as the normal white wood.

Durability and preservative treatment. Under damp conditions the timber is not resistant to fungal attack, and even in a dry situation it may suffer damage from the common furniture beetle in time. Ash is not particularly easy to treat with preservatives.

[*continued on page 16*]

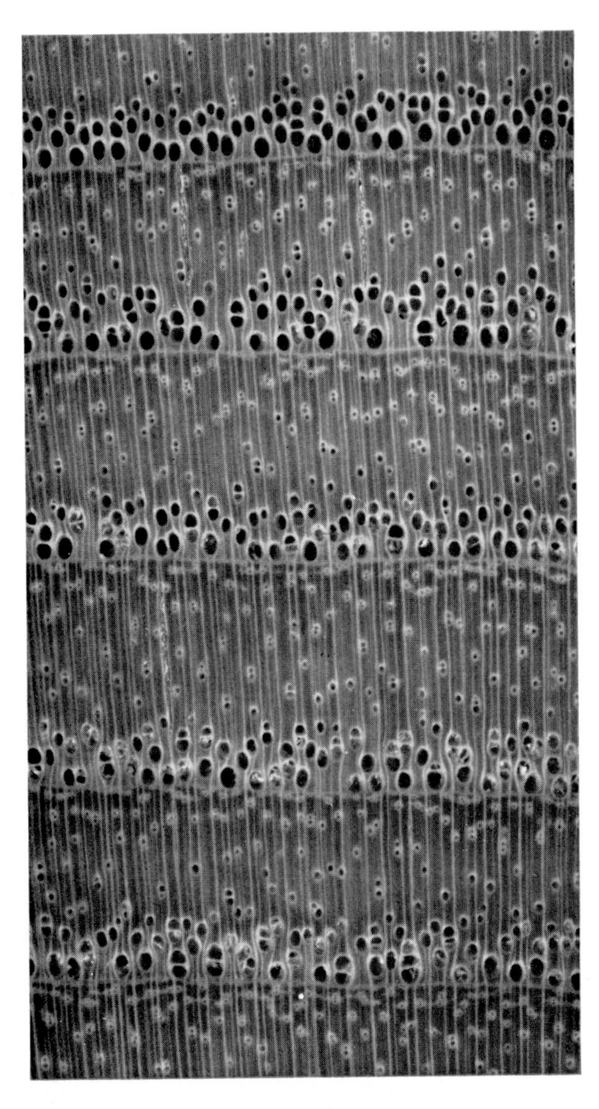

Fig. 1. Typical sample of fairly quickly grown ash with about 9 rings to the inch and an average of 70 per cent of summer wood.

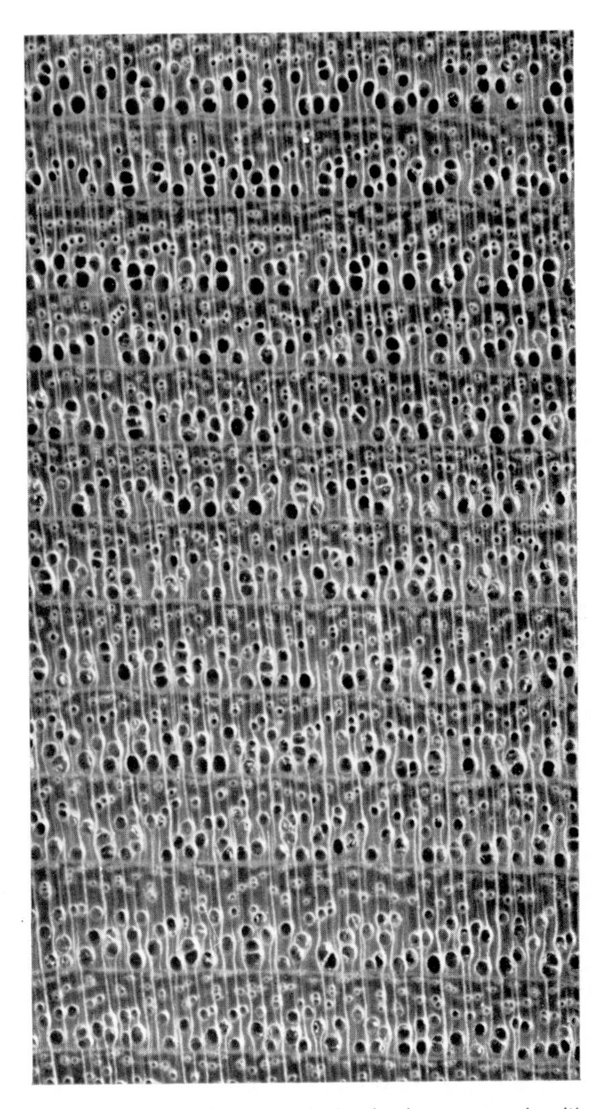

Fig. 2. Sample of comparatively slowly grown ash with an average of 22 rings to the inch and 40 per cent of summer wood.

Cross sections of ash showing variation in the structure of the wood. (Magnification x 10)

ASH [*continued from page 14*]

Working and finishing properties. In converting green timber there may be some binding on the saw and occasional splitting along the grain. Distortion in drying of flat-sawn timber may cause some waste in working, but after seasoning satisfactorily it is fairly easily worked and finished to a smooth finish.

Uses. Ash has many and varied uses; moreover it is one of the most difficult timbers to replace, either by another species of timber or by some different material. The highest quality timber—judged from the technical aspect—goes into the manufacture of sports goods, notably tennis rackets and hockey sticks, and the handles of striking tools, such as axes and hammers, and garden tools such as spades, rakes and hoes. In recent years the increasingly high performance demanded of sports goods has been obtained by laminating, using sawn or rotary-cut veneers. Considerable quantities of ash are used for motor lorry and bus bodies and for boat building. In the furniture industry the timber is used to a limited extent for special orders, and sometimes for the legs of oak furniture. Other typical uses are found in farm implements, agricultural machinery and gymnasium apparatus.

Selecting the timber by inspection. Ash is employed for highly exacting purposes and is often used to the limit of its strength. Recommendations for assessing the quality of the timber in the standing tree and in various stages of conversion are contained in two publications of the Forest Products Research Laboratory: Leaflet No. 47 *Selecting Ash by Inspection,* and Special Report No. 23 *The Quality of Ash from Different Parts of Britain.* The main conclusions are summarised below.
Differences in strength and bending properties between sites are small compared with the differences which may be found between the trees from any one site. It would appear that well-grown trees from any district favourable to the growth of ash can be expected to provide sports goods manufacturers with timber suited to their needs. The first consideration should be to choose trees with a good length of straight, clean bole, clear of side branches and with a well-developed, healthy crown at least one-half the height of the tree. Such trees are likely to be found in fairly open woodland; they should be preferred to tall, spindly trees with small crowns. Experience has shown that the best sports ash comes from trees between 0·3–0·4 m. (12–16 in.) in diameter, not more than 80 or 90 years of age, with fairly wide growth rings, say, between 2·5 and 4 rings to the centimetre (6–10 rings to the inch). Such trees generally have a smooth bark. Wood formed in later life tends to have narrower rings and may be deficient in strength. As a general rule, however, ash with between 1·5 and 6 rings to the centimetre (say, 4–15 rings to the inch) and a density of at least 0·58 (36 lb./ft.^{3}) in the seasoned condition will have reasonably good strength and bending properties, provided that the timber is free from deleterious knots, shakes, irregular grain, decay and insect attack.

Other species of interest. *American ash (the product of* F. americana *and allied species) is used for the same purposes as European. It is commonly graded as 'tough' (for tool handles, agricultural machinery and motor body work) and 'soft' (for cabinet work and interior joinery). Japanese ash (*F. mandshurica*), also known as tamo, is imported in the form of plywood.*

Ash

Flat-cut *Burr* *Quarter-cut*

Reproduced actual size

Beech

[*Fagus sylvatica*]

Distribution and supplies. Beech is one of the most important hardwoods of Western and Central Europe, including Britain. It grows to a height of 30 m. (100 ft) or more, with an average diameter of 1·2 m. (4 ft.). In close forest the smooth columnar bole may be clear of branches for 15 m. (50 ft.)—more often about 9 m. (30 ft.).
Beech is in good supply, in the form of logs and as square-edged and waney-edged boards, and is available in a wide range of sizes, 25–100mm. (1–4 in.) thick, 125–300 mm. (5–9 in.) wide, 2–3 m. ($6\frac{1}{2}$–10 ft.) long, also as 'shorts', dimension stock and, in semi-manufactured form, as plywood, flush doors and furniture parts.

General description. A general utility hardwood of plain appearance, usually straight grained and of fine, even texture. Timber grown in Britain is whitish or pale-brown, but much of the continental timber is steamed immediately after conversion—which gives it a warm reddish-brown colour. Darker-coloured, irregular markings ('rotkern' or 'red heart') are quite a common feature of Central European beech. Flat-sawn timber and rotary-cut plywood have practically no figure, though the rays, visible as dark lines or flecks against the lighter background, are characteristic. On quarter-sawn material, especially if the wood is given a natural finish, the rays are more prominent, producing a small, but nevertheless decorative, silver grain figure.
Beech grown in Britain, and timber imported from Northern Europe, is typically hard and dense, averaging about 0·72 (45 lb./ft.^{3}) in the seasoned condition; it is also lighter and more uniform in colour. By comparison, beech from Central Europe is generally milder and lighter in weight, of the order of 0·67 (42 lb./ft.^{3}), and, as noted above, a proportion of the timber is darker in colour, either as a result of the steaming treatment or from natural causes.

Seasoning and movement. Beech can be dried fairly quickly but is liable to check, split and warp in seasoning and the shrinkage is rather high. It also has a comparatively large dimensional movement in service.

Strength and bending properties. In the seasoned condition beech is superior to oak in bending strength, stiffness, hardness and resistance to impact and splitting. It is an exceptionally good wood for steam-bending purposes even when minor defects such as small knots and irregularities of the grain are present.

Durability and preservative treatment. In the natural condition the timber is not resistant to insect and fungal attack. Normally it responds well to preservative treatment by the hot and cold open-tank process, or under pressure, but the darker-coloured heartwood which commonly occurs in some continental beech is resistant to impregnation.

Working and finishing properties. In general, beech works fairly readily and finishes well in most hand and machine operations, particularly in turning. For mass production purposes, where ease of working is the main consideration, the milder wood from Central and Southern Europe is often preferred. Beech peels well and the cylindrical shape of the logs lends itself to rotary cutting for plywood. It can be glued without difficulty and can be stained to match with oak, mahogany or walnut.

[*continued on page 20*]

Beech

Flat-cut

Reproduced two-thirds actual size

BEECH [*continued from page 18*]

Uses. In Britain more beech is used than any other hardwood. Being plentiful and comparatively cheap, it has long been the standard timber where practical, rather than aesthetic, considerations determine the choice of material. It is the most widely used timber in the furniture industry, particularly for chairs. Its superior strength properties combined with moderate weight, good working qualities and clean, white appearance make it the preferred timber for brush backs, tool handles, parts of textile and other machinery, piano wrest planks, and a wide range of turned articles. In vehicle body building beech is used in laminated form for the wheel arches and other bent parts of caravans and for the framing of coach-built motor cars. Because of its low resistance to decay it is not usually considered suitable for structural purposes out of doors but there is no reason why it should not be used, after suitable preservative treatment, for bent work in boat building, e.g., as an alternative to rock elm and oak. As flooring, beech is suitable for heavy pedestrian traffic and for the light industrial type of factory.

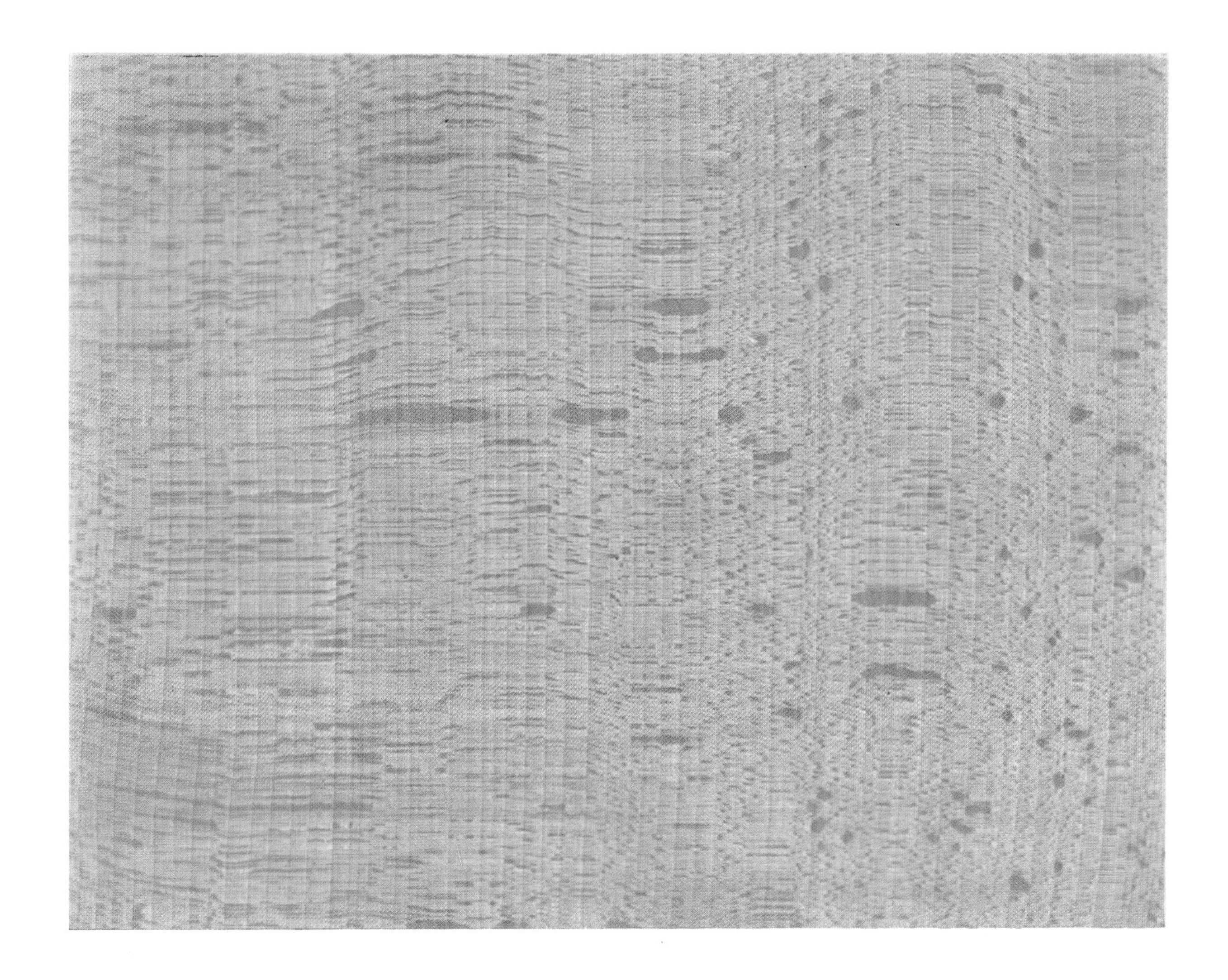

Beech

Quarter-cut

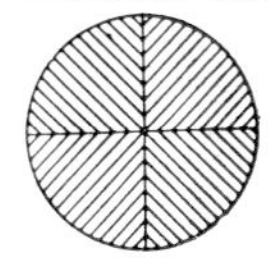

Reproduced actual size

Birch

[*Betula verrucosa* and *B. pubescens*]

Distribution and supplies. The two common species of European birch are widely distributed in Northern Europe, including the British Isles. So far as the timber is concerned there appears to be no significant difference between the species, and both grow naturally in open woodland, especially on poor sandy or peaty soil. They are relatively small trees, rarely exceeding 18 m. (60 ft.) in height. In Britain, where birch is not often encouraged by foresters, the trees are typically of poor form and the stem is apt to be fluted, but in Scandinavia, where it is regarded as an important timber tree, the bole is more cylindrical.

General description. Birch is normally a fairly straight-grained, fine-textured wood of plain appearance. Occasional trees have irregular grain, giving rise to a highly decorative figure. It has been found that logs with thin, smooth bark generally have straight-grained wood while in trees with rough bark the grain tends to be irregular (see Forest Products Research Laboratory Special Report No. 18 *Bark Form and Wood Figure in Home-Grown Birch*). The colour is normally white to light-brown throughout, with no distinct heartwood. The average density is about 0·66 (41 lb./ft.^{3}), seasoned.

Technical properties. Birch seasons fairly quickly, with some tendency to warp. When straight grained it is a good bending wood and is stronger than oak. It is not durable but is amenable to preservative treatment. It works fairly easily in most operations and takes a good finish if the grain is reasonably straight. The timber is excellent for turning, and gives good results with stain, polish and glue.

Uses. Ordinary straight-grained birch is useful as an inexpensive general utility wood having a plain appearance, high strength and good working and finishing properties. Being available in relatively small sizes (logs average 150 mm.—6 in.—diameter), it is largely used for turnery (bobbins, spools, cotton reels, brushes and brooms, small tool handles, etc.) and in the furniture industry for chair parts and upholstery frames. Most of the birch used for these purposes in Britain is imported, Sweden and Finland being major sources of supply. Large quantities of birch plywood are also imported into Britain from Northern Europe and the USSR. In recent years English birch has been recognised as a useful raw material for the pulp and paper industry.

In Scandinavia highly figured wood is used in the form of veneer for decorative work, and in the solid for fancy boxes, cigarette cases and the like, and there is no reason why selected English material should not be utilised in the same way. Karelian birch, masur birch and flame or ice birch are distinctively figured varieties.

Other species of interest. *Yellow birch from Canada and the USA is somewhat stronger, and denser and darker in colour than European birch, and is available in larger dimensions. The American white or paper birch resembles the European species more closely. Some Japanese birch is exported to Europe.*

Birch

Flat-cut

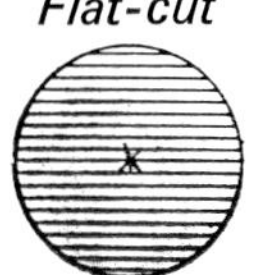

Reproduced two-thirds actual size

Boxwood

[*Buxus sempervirens*]

Buxus sempervirens is the original boxwood of Europe, Asia Minor and Asia, variously known as Turkey, Abassian, Iranian or Persian boxwood, etc., according to the place of origin. Certain other timbers which, though unrelated to true boxwood, are very similar in technical properties, are accepted as commercial boxwoods (see other species of interest below).

Distribution and supplies. Box is of limited occurrence, usually as a straggling shrub or small tree, up to 5 or 6 m. (say, 15 or 20 ft.) in height and rarely more than 150 mm. (6 in.) in diameter. It attains its best development in the Caucasus, from where it was formerly shipped as Abassian boxwood (from the Black Sea port of that name) in sizes of up to as much as 400 mm. (16 in.) in diameter. Nowadays true boxwood from various sources is obtainable in lengths of about 1 m. (3–4 ft.), 100–200 mm. (4–8 in.) in diameter.

General description. The characteristics of boxwood are its extremely fine, even texture and smooth, bright surface, its high density and clean light-yellow colour. The grain is sometimes straight but more often irregular. The density when seasoned is reported to vary from 0·83–1·14, average about 0·91 (52—71 lb./ft.^{3}, average about 57 lb.).

Technical properties. Boxwood is difficult to season and is liable to split badly if dried in the round; it should be dried very slowly. It is naturally rather hard to work but finishes well in most operations, cutting almost equally well in any direction of the grain. It has excellent turning properties both for plain and ornamental work. Boxwood is durable by reason of its high density but is not the type of timber normally used in conditions where resistance to decay is important.

Uses. Boxwood is the traditional material for wood-engravers' blocks, draughtsmen's rulers and scales, mallet heads, rollers and shuttles in the textile industry, parts of musical instruments, and a variety of small turned articles. To a large extent it has been superseded by other woods which are obtainable in larger sizes or at a lower price, or by modern synthetic materials.

Other species of interest. *The trade name boxwood covers the South African* Buxus macowani, *known as East London boxwood or Cape box, and a number of botanically unrelated species resembling true boxwood in general character. The more important of these are Knysna or kamassi boxwood (*Gonioma kamassi*) from South Africa, Maracaibo or West Indian boxwood (*Gossypiospermum praecox*) from Venezuela and Colombia, and San Domingo boxwood (*Phyllostylon brasiliensis*) from the West Indies.*

Boxwood

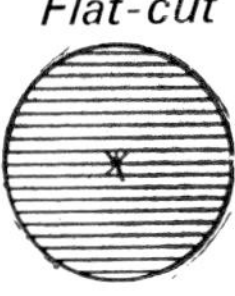

Reproduced actual size

Cherry

[principally *Prunus avium*]

Distribution and supplies. The wild cherry, or gean, frequently occurs with oak, beech and other hardwoods in mixed woodlands where it is conspicuous in spring for its white blossom. It may grow to a height of 18–25 m. (say 60–80 ft.) and a diameter of 0·6 m. (2 ft.) or more, large enough to yield timber for the sawmill. The cultivated cherry (*Prunus cerasus)* sometimes grows to a fairly large size in old orchards. Supplies of good quality cherry are strictly limited.

General description. Cherry-wood is of medium density (about 0·61 or 38 lb./ft.^{3} in the seasoned condition), with a fairly fine, even texture, generally straight grained except in the neighbourhood of large knots and near the butt. Flat-sawn material shows a well-marked growth-ring figure. The heartwood is light-pinkish or yellowish-brown, darkening in time to a mahogany shade (as in old furniture), but if the freshly worked wood is given a natural finish it acquires an attractive golden-brown colour, sometimes with a slightly greenish cast.

Technical properties. The timber seasons fairly readily but with a pronounced tendency to warp. Its movement in service is classed as medium. It is a tough timber, comparable to oak in strength, and is good material for steam bending—in the same class as beech, ash, oak and elm. Cherry is not sufficiently resistant to insect and fungal attack to be used out of doors without preservative treatment, and is liable to attack by the common furniture beetle. It works and finishes fairly well in most machine operations, turns well, gives excellent results with polishing, and presents no difficulty in staining and gluing.

Uses. Cherry is a high-grade cabinet timber which would probably be more highly appreciated and find greater use if supplies were more plentiful. In the form of furniture it is often seen in antique country-made pieces. Nowadays the limited supplies are used almost entirely for contract work. The good working qualities and attractive appearance of the timber make it suitable for interior joinery where large widths are not required. For panelling it is obtainable in the form of veneer.
A relatively large proportion of the timber available is of small dimensions, suitable for fancy goods and turnery, including wood-wind musical instruments such as recorders.

Other species of interest. *American black cherry (*P. serotina*) has been shipped to Europe from the USA.*
Cherry-wood tobacco pipes are made from Austrian cherry.

Cherry
Flat-cut

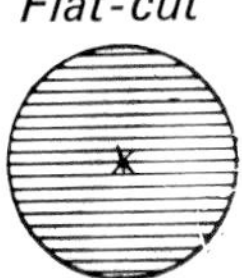

Quarter-cut

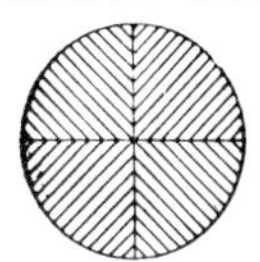

Reproduced actual size

Chestnut

[*Castanea sativa*]

Distribution and supplies. The sweet, or Spanish chestnut is supposed to have been introduced from Southern Europe by the Romans who cultivated the tree for its nuts. It has long been naturalised and is a familiar tree in the south and west of Britain where it commonly grows to a height of 30 m. (100 ft.) or more. The bole may be 6 m. (20 ft.) in length when grown in the open, and considerably more under forest conditions. Timber from mature trees tends to be spiral grained and is often affected by shakes (circular or radial cracks) which make it difficult to obtain sawn timber of any great width. For this reason it is generally grown on a short rotation of 60 years or so, or as coppice.
Cleft chestnut fencing and small quantities of sawn timber have been exported from France; the sawn timber is available in thicknesses of 25–75 mm. (1–3 in.), widths of 100–200 mm. (4–8 in.), in lengths of 2 m. ($6\frac{1}{2}$ ft.) and up.

General description. Chestnut is similar to oak in colour, grain and texture, and may easily be mistaken for plain oak at first glance, but owing to the absence of broad rays it lacks the characteristic silver grain of figured oak. Like oak, it has a corrosive effect on iron and steel and develops a blue-black discoloration in contact with these metals under damp conditions.

Technical properties. Chestnut is appreciably softer and lighter than oak, average density about 0·55 (34 lb./ft.^{3}) in the seasoned condition, and it is correspondingly inferior in most strength properties. It seasons slowly and unevenly with a tendency to collapse and honeycomb but when thoroughly seasoned the timber is more stable than oak under varying conditions of humidity. It is suitable for bent work. In resistance to decay chestnut is almost as good as oak, and has the advantage that the sapwood is exceptionally narrow, rarely exceeding $\frac{1}{2}$ in. It is fairly easy to work and takes a good finish.

Uses. From its resemblance to plain oak, chestnut of suitable dimensions and quality is used as an alternative to oak where only moderate strength is required, as in furniture—especially for the legs of chairs and tables—and for coffins and sills. Its principal use is for cleft fencing, made from coppice growth which is typically straight grained, and is easily cleft. Large quantities of coppice chestnut were formerly used for hop-poles but the increasing use of iron has reduced the demand. Its high natural durability makes chestnut a valuable timber for estate work.

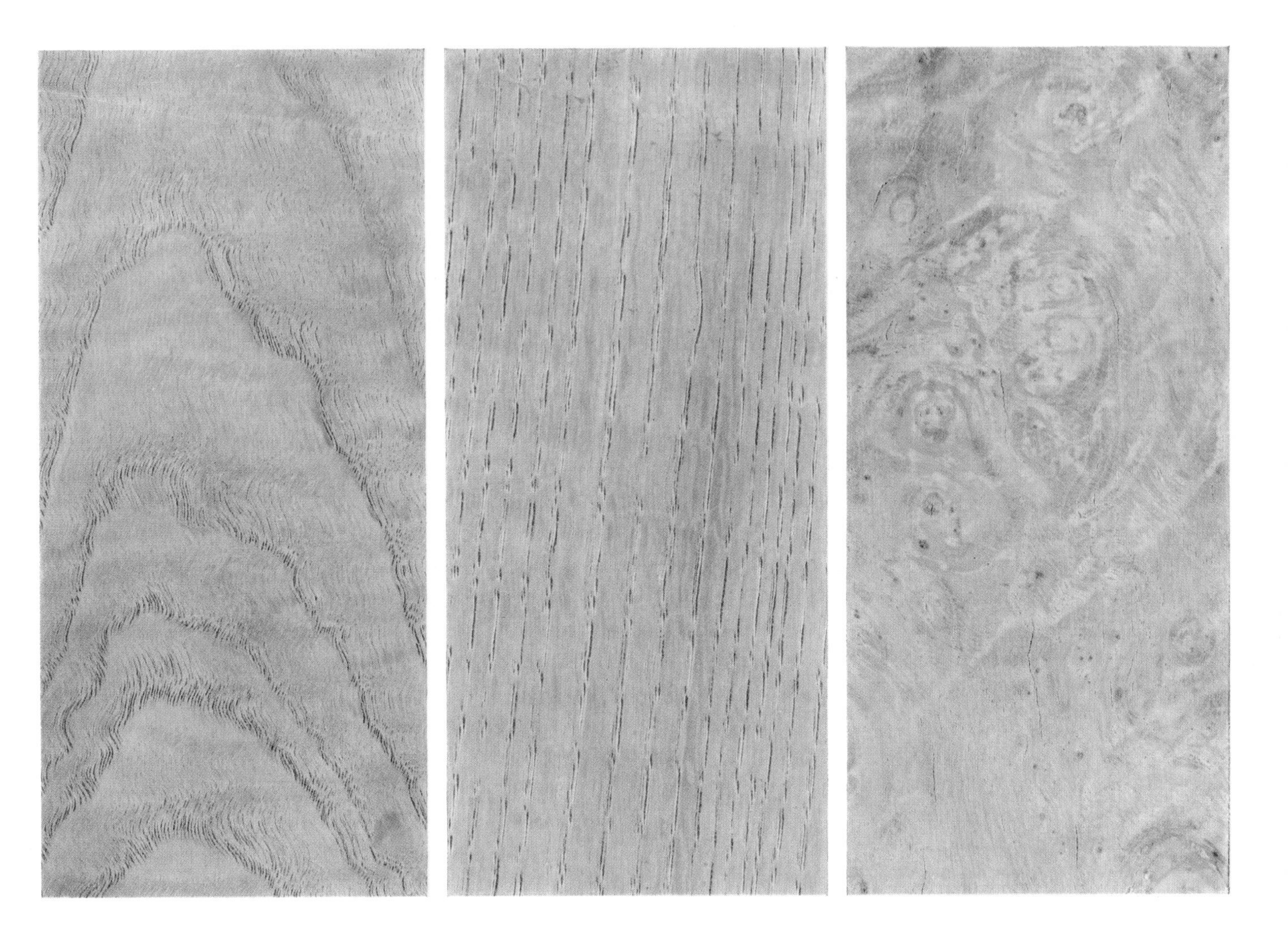

Chestnut

Flat-cut

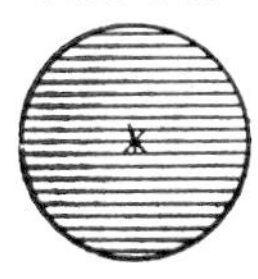

Quarter-cut

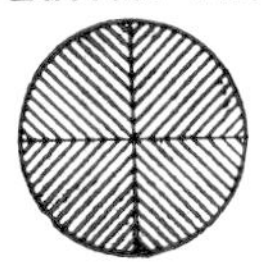

Burr

Reproduced actual size

Elm

[various species of *Ulmus*]

Distribution and supplies. The principal species in the south of England is the English elm (*Ulmus procera*), usually grown as a hedgerow tree. In other parts of the country this is replaced by local varieties—Dutch elm, Huntingdon elm, Cornish elm, wych elm or Scotch elm, etc. Once the logs are converted the different kinds of elm are not usually distinguished commercially, except for the wych elm (*U. glabra*) which is technically superior to the others, as noted below. Hedgerow elms commonly grow to a very large size, 37–46 m. (120–150 ft.) in height and up to 1·5 m. (5 ft.), or even 2·4 m. (8 ft.) in diameter. The bole of the English elm may be more or less clear for 12 or 15 m. (40 or 50 ft.). The larger logs are often hollow or rotten in the heart. Sawn timber is available in a wide range of sizes.

General description. In the green condition the light-coloured sapwood of English elm is sharply distinct from the reddish-brown heartwood; on drying the heartwood tones down to a dull brown. The large pores give the wood an open texture. Largely due to the way the trees are grown, with heavy spreading branches, most elm timber is rough and uneven with irregular grain (cross grained). Wych elm, however, is typically straighter in the grain and not so coarse in texture; the heartwood is light brown, often with a greenish tinge or distinct green streaks.
Elm imported from the Continent is likewise typically straighter in the grain, more evenly grown and plainer in appearance than the general run of English elm.

Technical properties. English elm is fairly light in weight—density about 0·55 (34 lb./ft.^{3}) in the dry condition—appreciably lighter than oak and correspondingly inferior in most strength properties. Its reputation for toughness is due to the characteristically interlocked grain which makes it difficult to split. The cross-grained character of the wood also accounts for its reputation for distorting and splitting. With modern techniques of kiln seasoning it is possible to reduce this form of degrade to a minimum; nevertheless elm is not the sort of timber that should be chosen for purposes where stability under varying conditions of humidity is important. It is somewhat troublesome in conversion, as the wild grain tends to cause binding on the saw. Planing to a smooth finish is often complicated by 'picking up' or tearing of the surface where the grain is twisted or wavy. In other machining operations there is little difficulty provided that the tools are in good condition. Elm takes nails without splitting, and glued joints hold satisfactorily. For cabinet work, panelling and similar purposes, scraping gives a good finish.
Elm is a good wood for steam bending, being comparable to beech and ash in this respect, but only if it is carefully selected for freedom from defects such as knots and irregular grain.
Wych elm is appreciably heavier—average about 42 lb./ft.^{3}—and stronger than most other varieties of elm. It is almost as tough as ash, and is a useful substitute for ash for steam-bending work. Being fairly straight grained it gives little trouble in sawing, and has better working properties than the general run of English elm.
Elm, generally, is not durable under conditions favourable to decay. For use out of doors it should be given a preservative treatment.

Uses. Elm has a wide range of uses since it is a cheap hardwood, strong and tough for its weight, and readily obtainable in large sizes. Where supplies are available it is commonly used in the form of waney-edged boards for weather

[*continued on page 32*]

Elm

Flat-cut

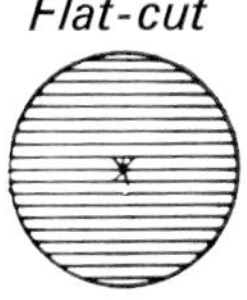

Reproduced two-thirds actual size

ELM [*continued from page 30*]

boarding on farm buildings, and for rough constructional work such as mangers, stall divisions, cattle cribs, etc. Treated with creosote it is useful for most estate purposes. It is a traditional timber for cheap coffins, for country-made furniture, such as refectory tables, garden furniture and the seats of Windsor chairs, and can also be used for high-class furniture and cabinet work provided that it is given an appropriate seasoning and conditioning treatment. The grain is shown to advantage, especially when polished, in turnery products such as fruit and salad bowls.
In boat building elm is used principally for keels, rudders and rubbing strakes, and sometimes for planking. Wych elm is preferred because it is straighter in the grain, tougher, easier to work and less liable to distort. If it is kept wet, particularly in brackish water, it lasts very well but is not recommended for the keels or planking of boats sailing in fresh water, and likely to be laid up under conditions favouring fungal decay. In dock and harbour work elm is commonly used for fenders.
Because it holds nails well it is suitable for the ends of heavy-duty boxes and crates.
Decorative veneer for panelling and furniture is cut from selected logs and from the highly figured wood of burrs.

Other species of interest. *Rock elm (*U. thomasii*), of Canada and the USA, is a special-purpose timber used mainly for ship building and agricultural machinery. Japanese elm or nire (*U. davidiana *and other species) has been shipped to Europe in small quantities, mainly as square-edged boards and strips.*

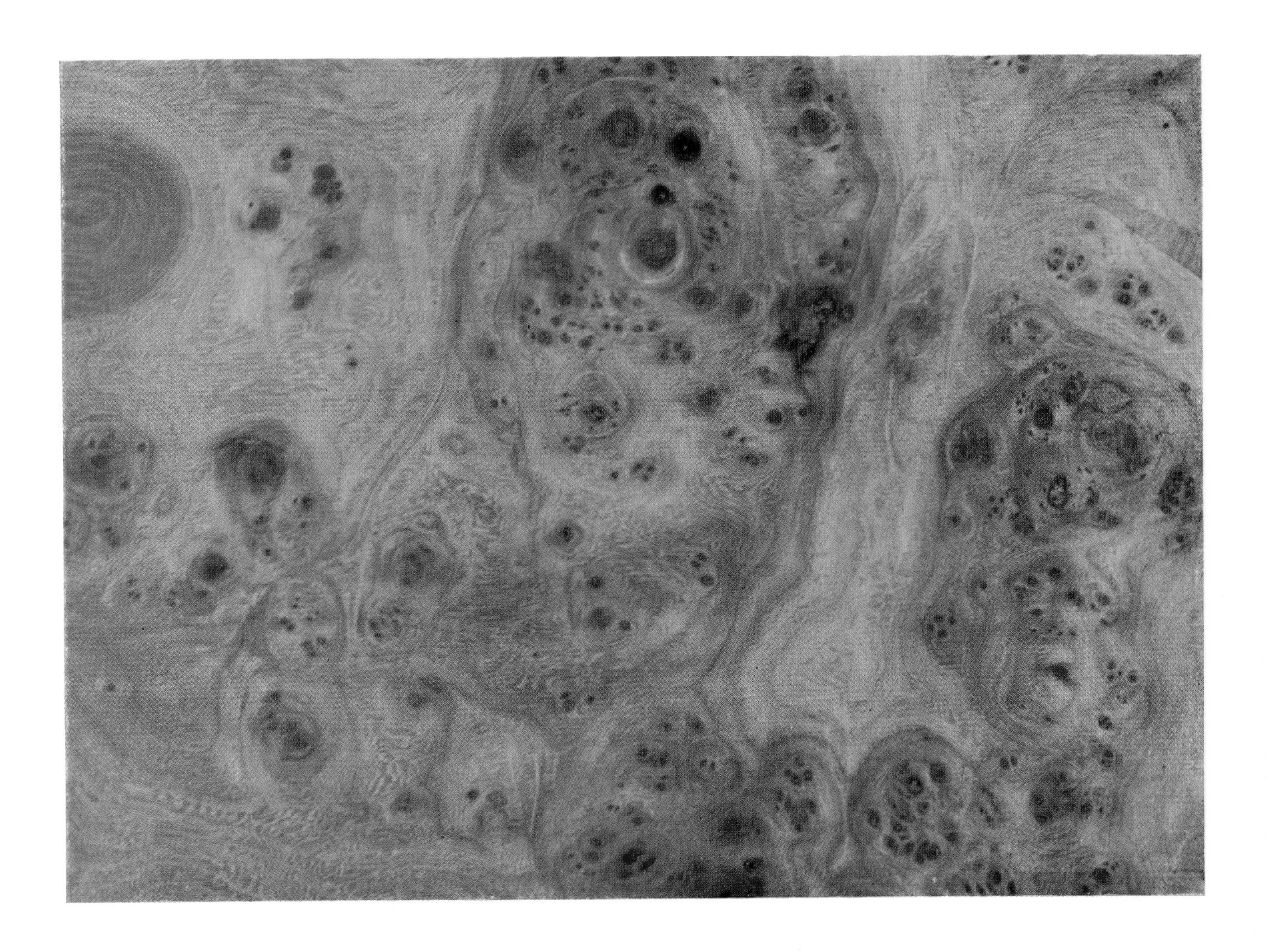

Elm

Burr

Reproduced two-thirds actual size

CO

Holly

[*Ilex aquifolium*]

Distribution and supplies. Holly is not usually considered a commercial timber but is of sufficient interest to be included. It is a common tree of the woods and hedgerows, seldom more than 9 m. (30 ft.) high and 0·3 m. (1 ft.) in diameter, though larger specimen trees are found in parks and gardens.

General description. An unusually dense wood (about 0·80 or 50 lb./ft.^{3}, seasoned) with a fine, even texture, dull white with a greenish or grey tinge. The grain is inclined to be irregular.

Technical properties. A difficult timber to dry; it is apt to warp badly unless cut to small sizes. Relatively hard to work but capable of an excellent finish, and takes stains well. It is not resistant to decay.

Uses. Supplies of good-quality timber are limited. The logs are usually small, irregular and knotty, so the wood can only be used in small sizes. Its main uses are for inlay work and marquetry, small fancy articles, wood engraving, turnery and—when stained black—as a substitute for ebony.

Holly

Flat-cut

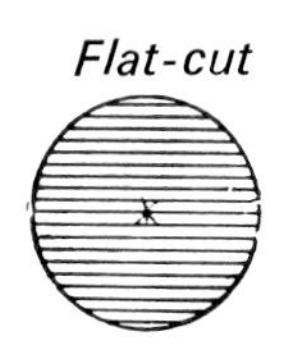

Reproduced actual size

Hornbeam

[*Carpinus betulus*]

Distribution and supplies. Hornbeam is widely distributed throughout Europe, though nowhere very abundant. Hornbeam is a medium-sized tree, seldom exceeding 18 m. (60 ft.) in height and 0.6 m. (2 ft.) in diameter, somewhat resembling beech but distinguished by the deeply-fluted trunk which usually branches fairly low down, causing considerable waste in conversion; the timber is not therefore available in large sizes.

General description. A dense wood with a fine, even texture somewhat similar to beech in general appearance but dull white in colour with greyish streaks (cf. the German *Weissbuche*, literally 'white beech'). The fluted shape of the trunk is reflected in the undulating course of the annual rings on the end grain. The timber is typically cross grained which makes it tough and difficult to split. The average density is about 0·75 (47 lb./ft.^{3}), slightly denser than beech and oak.

Technical properties. Hornbeam is outstanding among European hardwoods for its combination of high density, strength, toughness and resistance to splitting. It is also classed as very good for steam bending. Though hard, it can be sawn fairly readily in the green condition. Hornbeam is rather difficult to work in the dry state, being similar to a dense grade of beech in this respect. It finishes very smoothly, turns well, holds screws firmly and gives excellent results with stain and polish. It presents no difficulty in seasoning and is fairly stable in service. The timber is not durable under conditions favourable to decay but, being permeable, is amenable to preservative treatment and impregnation with dyes.

Uses. Because of its exceptional toughness, its resistance to splitting and its good working and finishing qualities, hornbeam was formerly in demand for cogs and working parts of wooden machinery, wheelwrights' work, wood screws, pegs and mallet heads. Probably its principal industrial application now is for the small parts of piano actions. The advantage of hornbeam for these complicated items is that it takes a good finish from a very fine saw, can be bored with small, closely spaced holes, holds screws firmly and remains fairly stable (swelling would cause jamming of the action). Most of the hornbeam used for piano work originates in France. French hornbeam is also used for the shafts of billiard cues, as an alternative to English ash.
Hornbeam dyed black is sometimes used as a substitute for ebony.
As flooring it has a high resistance to abrasive action and wears comparatively smoothly; it may be considered as an alternative to maple.

Hornbeam

Flat-cut

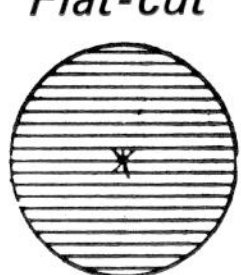

Reproduced two-thirds actual size

Horse Chestnut

[*Aesculus hippocastanum*]

Distribution and supplies. A native of Greece and parts of Asia, the horse chestnut is very commonly grown for shade or ornamental purposes. It has not been considered a suitable tree for forestry because the timber is of little value; however, its rapid growth suggests that it may have possibilities for pulp or veneer production. It grows to a large size, 30 m. (100 ft.) or more in height and 1·5 m. (5 ft.) in diameter. Horse chestnut is related to the American buckeye.

General description. A soft, light-weight hardwood, density about 0·51 (32 lb./ft^3.), about the same as yellow deal. Creamy-white or yellowish with a fine, even texture, resembling lime, willow and poplar in appearance; inclined to be cross grained.

Technical properties. The timber dries quickly with little degrade, and does not shrink or swell much in service. It works readily and takes a clean, smooth finish—provided that sharp, thin-edged tools are used—and gives good results with stain, polish and glue. Horse chestnut is soft and brittle and is susceptible to fungal and insect attack but readily absorbs wood preservatives.

Uses. Horse chestnut is of little interest to the timber trade; the limited supplies are of some value, however, on account of its clean, white appearance and good working properties, for such purposes as kitchen and dairy utensils, fruit storage trays, boxes and crates for fruit and vegetables, toy making and general turnery. It is used as an alternative to sycamore and lime for the neck wedges of tennis rackets and for the facings on handles.
Rotary-cut veneer has been used for making chip baskets for holding fruit and vegetables. The timber is suitable for carving.

Horse Chestnut

Flat-cut

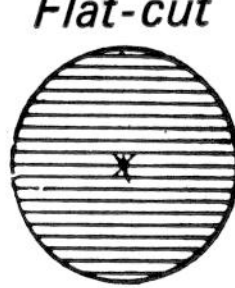

Reproduced actual size

Laburnum

[principally *Laburnum anagyroides*]

Distribution and supplies. The common laburnum (*L. anagyroides*) and the so-called Scotch laburnum (*L. alpinum*) are ornamental trees introduced from Central and Southern Europe, and widely cultivated in parks and gardens and on roadsides. The trees seldom exceed 9 m. (30 ft.) in height and 250 mm. (10 in.) in diameter, and the trunk tends to branch low down, so they are of little interest to the timber trade.

General description. In general appearance and in technical properties laburnum resembles robinia. A cross-section of the trunk shows a narrow band of yellowish sapwood sharply contrasting with the golden-brown heartwood. On exposure to light and air the heartwood darkens to a deep-brown, relieved by darker brown or nearly black markings. The machined surface has a natural lustre which helps to give the wood a highly decorative appearance. It has been called false ebony. Laburnum is one of the hardest and heaviest of European timbers, density 0·80–0·88 (50–55 lb./ft.^{3}) in the seasoned condition. It is generally straight grained, with a fine texture which appears coarser than it really is because of the contrast between the light-coloured rays and zig-zag lines of soft tissue and the dark background of dense fibrous tissue.

Technical properties. Although the timber has not been subjected to systematic tests, it has the reputation of being durable, suitable for use in the ground without preservative treatment. It is said to be easy to season. Though hard, it works fairly readily with sharp tools, taking a smooth bright finish. It is particularly good for turning.

Uses. At one time laburnum was used in the solid and in the form of veneer for cabinet work and inlay, for musical instruments such as recorders, flutes and bagpipes, for knife handles and ornamental turnery. It can be seen in period furniture in the form of 'oysters' (end-grain veneers). Very little use is made of the timber today, though if supplies were organised there appears to be no reason why this beautiful wood should not compare favourably with such timbers as rosewood, ebony, partridgewood and cocuswood—all imported at great cost—for purposes where small-dimension decorative hardwood is required.

Laburnum

Quarter-cut

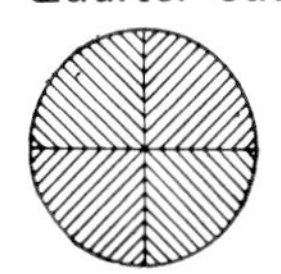

Reproduced actual size

Lime

[principally *Tilia vulgaris*]

Distribution and supplies. Lime is of more interest as an ornamental tree than as a source of commercial timber. It occurs sporadically throughout Europe, including Britain, in mixed woodlands, but is not often used in plantation work. It is a large, spreading tree, commonly 30 m. (100 ft.) or more high and up to 1·2 m (4 ft.) in diameter. The trunk is often disfigured by large burrs. Since it is usually grown in the open—in avenues, parks and gardens—the timber is mostly of poor quality and not much in demand, but it can generally be obtained through timber merchants if required.

General description. The wood is of plain appearance, soft but firm with a fine, even texture, straight grained when clear of knots and other irregularities; nearly white or pale-yellow, turning pale-brown on exposure. The density is about 0·55 (34 lb./ft.^{3}) in the seasoned condition.

Technical properties. It seasons well and rapidly, with some tendency to distort. The movement of the wood in service is about average. It is fairly strong for its weight and resistant to splitting but softer than the general run of hardwoods. Lime is not a good bending wood. It is perishable but amenable to preservative treatment. It works easily and takes a clean, smooth finish provided that sharp, thin-edged tools are used.

Uses. European lime was formerly used for a variety of purposes calling for a fairly soft, light-weight hardwood of clean appearance, e.g., in toys, handles for paint brushes, turnery, hat blocks, frames for beehives, parts of tennis rackets; also for piano keys. It has been largely replaced by the North American lime (generally known as basswood) and Japanese lime (shina) and tropical light hardwoods such as obeche, which are more consistent in quality.
Lime is considered one of the best European timbers for carving, as it is fairly soft and cuts with a clean, smooth surface in all directions of the grain.

Lime

Flat-cut

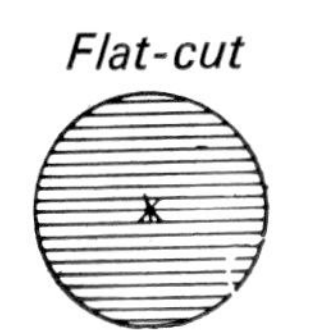

Reproduced actual size

Oak

[*Quercus robur* and *Q. petraea*]

English or European oak is obtained from two closely allied botanical species, the common English or pedunculate oak (*Quercus robur*) and the sessile or durmast oak (*Q. petraea*). So far as the timber is concerned there appears to be no significant difference between them, though the sessile oak typically has a longer, straighter trunk than the pedunculate.

Distribution and supplies. Oak is widely distributed in Western and Central Europe; major sources of supply are France, Poland, Yugoslavia and the Baltic countries. It is the most common forest tree in Britain, especially in England and Wales; both species are forest trees which, when well grown, have long, straight stems and fairly compact crowns, and it is from such trees that the best English oak timber is obtained. However, much oak woodland has an open character, with large trees, branching low down and developing huge crowns; in the hedgerows, too, oak is often of this form and it is from these trees that much of the lower-grade timber is obtained.
Sawn timber is obtainable from stock in a wide range of sizes from 16–150 mm. ($\frac{5}{8}$–6 in.) thick, 125 mm. (5 in.) and up in width, 2–3 m. ($6\frac{1}{2}$–10 ft.) or more in length, square-edged and waney-edged, and in shorts, strips and squares, also as veneers for decorative work. Much of the lower-grade timber is sawn into posts and rails for fencing, gates, and coffin boards; logs are always available for conversion to special sizes for structural and other uses.

General description. Oak heartwood, when freshly cut, is normally pale yellowish-brown in colour; on exposure to light it assumes a warm brown shade and becomes darker with age. (A special form, known as brown oak, is described below.) The light-coloured sapwood, 25–50 mm. (1–2 in.) wide, is quite distinct. The timber is extremely variable in character, largely due to the influence of growth conditions on the structure of the wood. When growth is rapid, as in vigorous young trees, the rings are wide with a large proportion of dense summer wood, and the timber is consequently hard and heavy and well adapted for constructional purposes. By contrast, wood that is grown slowly—as in old age or when the trees are closely spaced so that development is restricted—has narrow rings, and is comparatively soft, light in weight, and more suitable for conversion to veneer, or for the manufacture of furniture or high-class joinery.
Timber produced in Central Europe, such as Slavonian oak from Yugoslavia, is typically of slow, even growth, mild and easy to work, with an average density of about 0·67 (42 lb./ft.^{3}), seasoned, while oak from France, Germany and Northern Poland tends to be somewhat harder and tougher. The general run of English oak is comparatively hard and dense, averaging about 0·72 (45 lb./ft.^{3}), seasoned. However, selected material from well-grown trees compares favourably with corresponding grades of imported timber for cabinet work and joinery, while the slowly grown, narrow-ringed wood from old oak butts is excellent for decorative veneers.
The ornamental silver-grain figure of quarter-sawn (radially cut) oak, is characteristic, due to its broad rays. On fully quartered material the rays are clearly visible as lustrous, discontinuous bands or flakes. When cut at a slight angle to the radius the figure is broken and less conspicuous. Flat-sawn timber (known as plain oak) and rotary-cut veneer show a well-marked growth ring figure, not unlike that of ash, elm, chestnut and other ring-porous hardwoods, the rays being just visible as long, slender spindle-shaped lines. Oak is usually straight grained except in the vicinity of knots.

[*continued on page 46*]

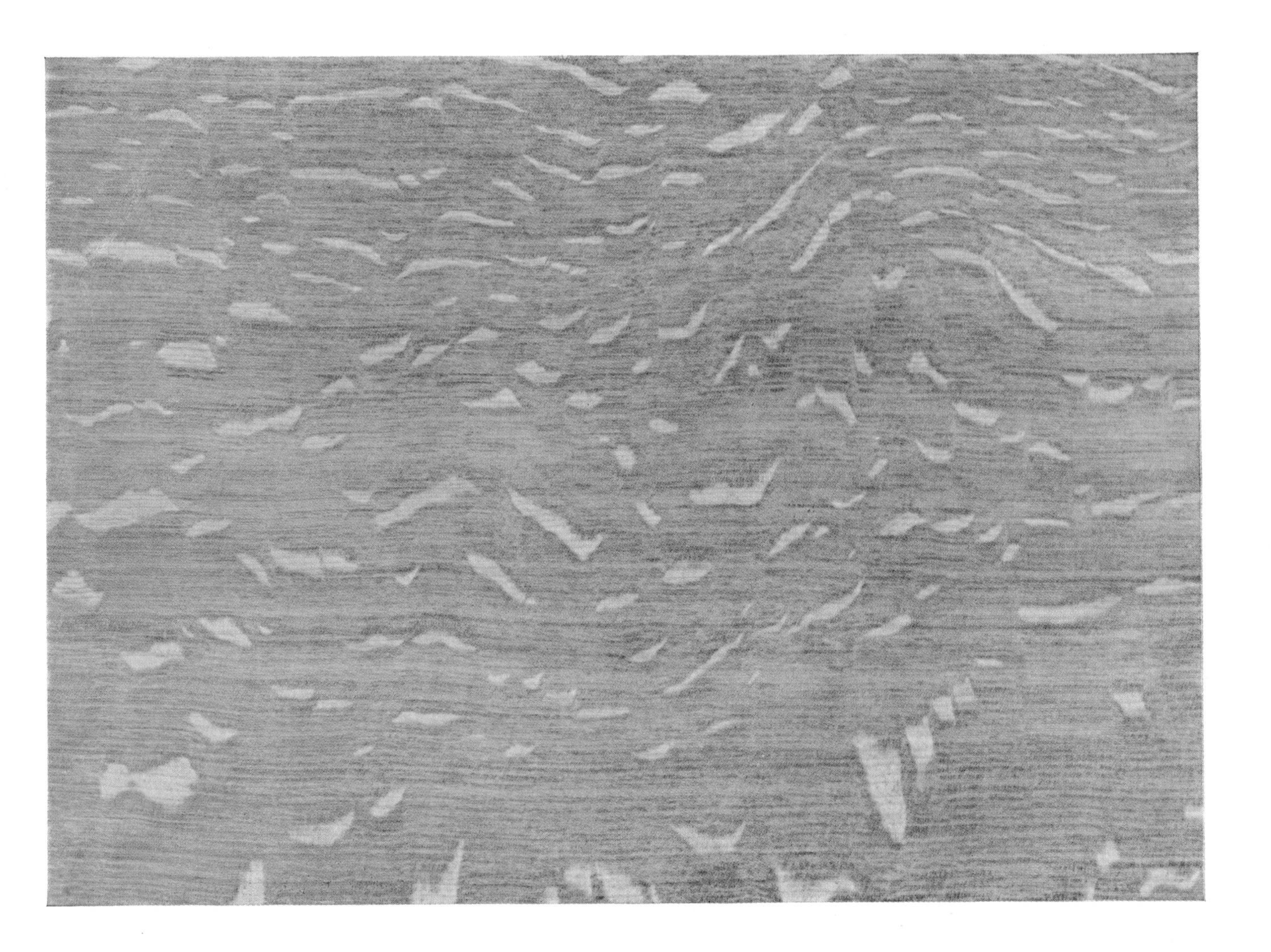

Oak

Quarter-cut

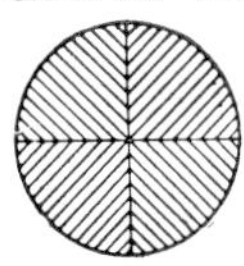

Reproduced two-thirds actual size

OAK [*continued from page 44*]

The wood has a corrosive action on metals, especially iron, steel and lead, particularly if the wood is moist. Blue-black discolorations, due to a reaction between iron and the tannin in the wood, are liable to develop on the surface of oak in contact with iron or iron compounds under damp conditions. The use of non-ferrous metals for fastenings and fittings is recommended where practicable.

Seasoning and movement. Oak is not an easy timber to dry, and under outdoor conditions the inside of thick-dimension stock may remain wet for many years; kiln drying, too, is a slow process, and considerable care is needed to avoid excessive degrade. The wood has an appreciable shrinkage on drying, and when dry its movement in service is rated as medium.

Strength and bending properties. The strength properties of oak are well known, in fact it is commonly taken as the standard to which other timbers are compared. It is also a very good wood for steam bending.

Durability and preservative treatment. Oak has long been accepted as synonymous with durability. The sapwood, however, is not durable and should be excluded from timber intended for indoor use because of its susceptibility to insect attack. For purposes such as fencing and constructional work where sapwood is not cut away, effective treatment with a wood preservative will add appreciably to the life of the timber.

Working and finishing properties. In general, oak can be sawn and finished fairly readily, though heavy, tough timber may present some difficulty in working; also, the quality of the finish depends on the presence of knots and other defects which cause picking up, although their effect can be mitigated by reducing cutting angles to 20°.

Uses. Because it is strong and durable and in good supply, oak is the traditional timber for high-grade constructional work in most European countries. Large quantities are used in ship and boat building, mainly for keels and framing. Curved parts for the framing of boats, which were formerly made from branchwood, are being replaced by laminated bends made by bonding together thin boards of oak with resorcinol glue. Lower-grade material is also widely used for estate purposes, fence posts, gates, motorway fencing and the construction of heavy road vehicles. Oak retains its popularity where aesthetic considerations carry weight, notably for interior woodwork in public buildings, for exterior joinery in conventional house building and for flooring in the form of blocks and strips. Another traditional usage is for coffins. It is out of fashion for contemporary domestic furniture but is always in demand for contract work. An important specialised use is in the manufacture of casks for beer, wine and spirits, and vats and tanks for industrial processes.

Brown oak. The heartwood of old trees is sometimes streaked or uniformly coloured a distinctive, warm brown due to the action of a fungus, the so-called beefsteak or liver fungus (*Fistulina hepatica*). The fungus dies when the timber is dried so there is no danger of its causing decay after conversion, and although the strength of the timber is somewhat reduced this is not important for purposes such as panelling, furniture, interior fittings and fancy goods. It is particularly popular in the USA.

Other species of interest. *The deciduous oaks of North America and Northern Asia produce timber similar to English or European oak in general character. The more important commercial varieties are American white and red oak, Japanese oak and Persian oak. The evergreen oaks, which grow mainly in subtropical and tropical climates, are of inferior quality so far as their timber is concerned.*
*The name oak is also applied to certain other timbers with some resemblance to oak. Australian silky oak (*Cardwellia sublimis*) is a decorative timber used mainly in the form of veneer for panelling, etc. Tasmanian oak (*Eucalyptus *species) resembles plain oak in grain and texture.*

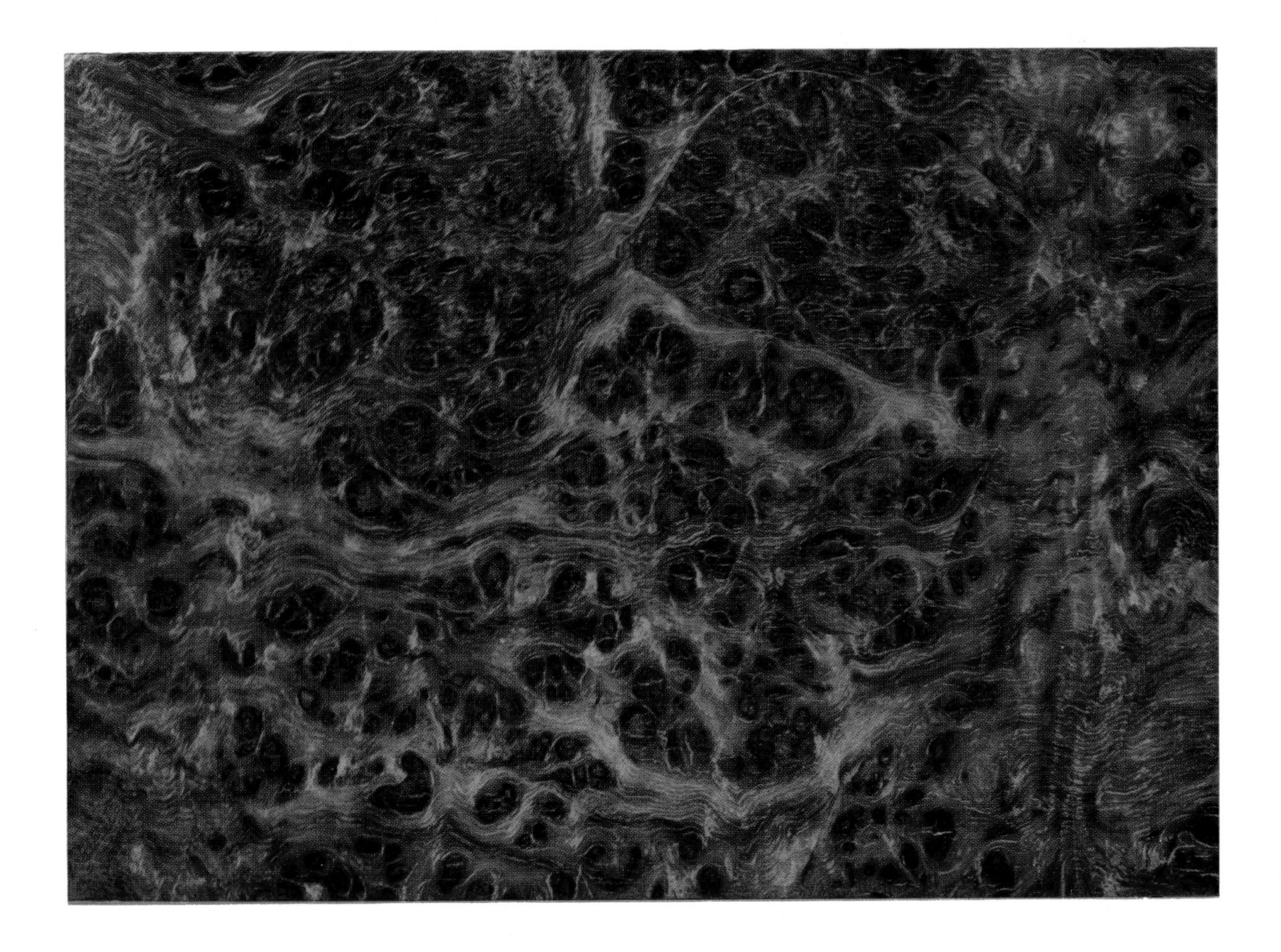

Oak

Burr

Reproduced actual size

Pear

[*Pyrus communis*]

Distribution and supplies. The wild pear of Europe and Northern Asia is the parent of the numerous cultivated varieties. It is usually a small to medium-sized tree, up to 9–12 m. (30–40 ft.) in height and 0·3–0·45 m. (12–18 in.) in diameter. Most of the timber that reaches the market probably comes from old orchards.

General description. Well-grown trees yield timber of excellent quality, dense—about 0·72 (45 lb./ft.^{3}) in the seasoned condition—with a fine, uniform texture, pinkish-brown, usually with no distinct heartwood, though the central core of old trees is sometimes dark brown. The grain is inclined to be irregular; occasional logs yield figured wood suitable for decorative veneer.

Technical properties. The timber must be dried slowly to minimise warping; when properly seasoned and kept dry it holds its shape well enough to be used for precision work. It is hard to work but cuts cleanly in all directions of the grain, finishing with a smooth, but fairly dull surface. The wood turns well and is excellent for carving. It is said to be similar to oak in most strength properties but tougher and more difficult to split. It is not sufficiently durable to be used outdoors without preservative treatment.

Uses. Being hard, fine-textured and light in colour, pear wood was formerly the standard material for drawing instruments such as set-squares and curves. It is still in demand for wood engraving, calico printing blocks, tool handles, especially hand-saws, and for turnery. Stained black, the wood is employed as a substitute for ebony in certain musical instruments and for picture frames and small handles of various kinds. Many of the recorders used in music classes in schools are made of pear wood. Other exacting uses are for the shafts of billiard cues and for carving. Pear is probably used more widely on the Continent than in Britain.

Other species of interest. *The wood of apple (*Malus sylvestris*) resembles pear in structure and general properties. It is generally considered inferior to pear, being more irregular in the grain, and variable in colour. It may be used as a substitute for pear and for the heads of masons' and carpenters' mallets.*
*Several other woods of the same family have similar characteristics to apple and pear, but owing to their small size, or scarcity, are of little economic importance. Among them are the rowan or mountain ash (*Sorbus aucuparia*), the whitebeam (*S. aria*) and the service trees (*S. domestica *and* S. torminalis*). The hawthorn (*Crataegus oxyacantha*) produces a very hard, heavy wood, darker in colour than pear or apple but essentially similar in structure. The trunk is typically irregular, often twisted and deeply furrowed, and is in consequence of comparatively little value as timber, though it is eminently suitable for the heads of heavy mallets.*

Pear

Flat-cut

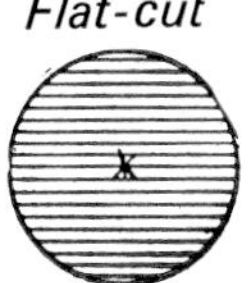

Reproduced actual size

DO

Plane

[principally *Platanus acerifolia*]

Distribution and supplies. Two species of plane are commonly grown in Britain as ornamental trees, namely the Oriental plane (*Platanus orientalis*), of South-East Europe and Asia Minor, and the London plane (*P. acerifolia*), supposed to be a cross between the Oriental plane and the American western plane (*P. occidentalis*), commonly known as sycamore in the USA. The London plane is a favourite species for planting in towns and cities, where it flourishes in spite of the smoky atmosphere and attains magnificent proportions, 30 m. (100 ft.) in height and 1 m. (say, 3–4 ft.), or more, in diameter with a clear bole of 10 m. (30 ft. or more). It is seldom, if ever, planted for timber production, and practically the only source of timber is specimen trees. Supplies are limited and it is not regularly stocked by timber merchants but can generally be obtained on enquiry.

General description. In colour and texture the wood resembles beech, but is readily distinguished by the numerous deep rays, which give a distinctive and highly decorative figure to quartered timber. They are of reddish-brown colour and are very conspicuous against the lighter-coloured background tissue. Timber specially cut to display this ray figure is often termed lacewood and is employed for fancy articles and panelling, chiefly in the form of veneer. Generally there is no marked difference between sapwood and heartwood, but some logs, when opened, reveal an irregular core of darker wood.

Technical properties. Plane is relatively softer and lighter than beech, density 0·64 (about 40 lb./ft.^{3}) in the seasoned condition. It is not quite so strong as beech, perishable, but fairly easy to work in most hand and machine operations. The tendency for the rays to break away under the tool when quartered material is machined can be minimised by a slight reduction of the cutting angle.

Uses. The limited supplies of plane are used mainly for decorative veneer, sliced so as to show the unique ray figure. Veneers are used in interior decoration for panelling, for cabinet work, fancy goods and decorative work.

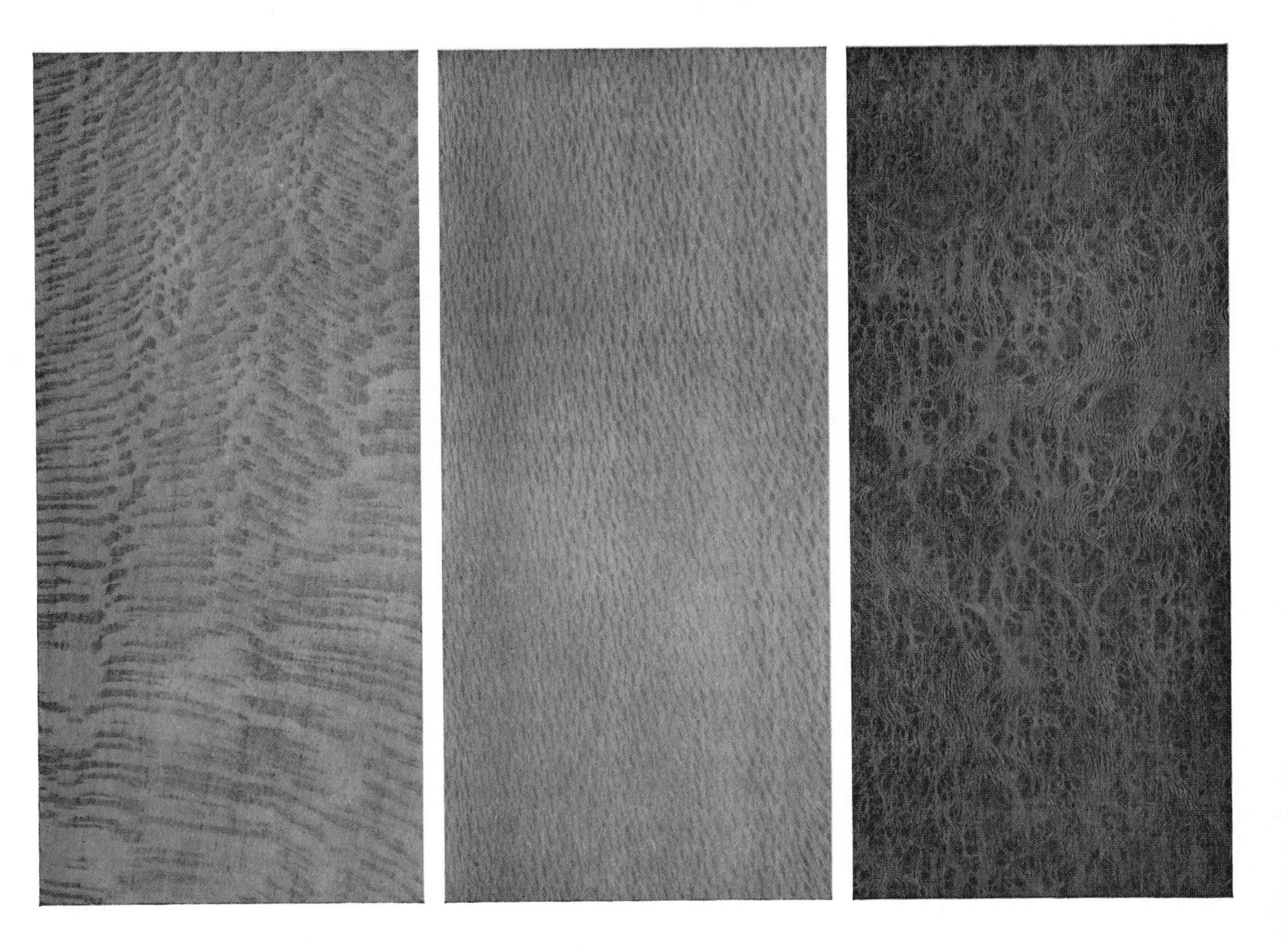

Plane

Quarter-cut

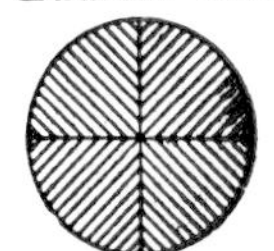

Reproduced actual size

Quarter-cut

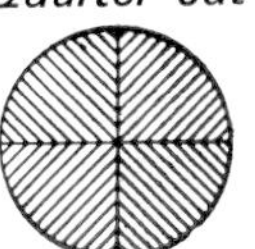

Burr

Poplar

[various species of *Populus*]

The poplars fall naturally into four main groups: (*a*) balsam poplars, (*b*) white poplars, (*c*) aspens, and (*d*) black poplars. The last two groups are important in timber production.
It should be noted that in the USA the name poplar or yellow poplar refers to *Liriodendron tulipifera,* better known to the timber trade in Britain as American whitewood or Canary whitewood. It is occasionally grown in Britain and is known as the tulip tree. In Canada and the USA true poplar is commonly known as cottonwood.

Distribution and supplies. The European aspen (*Populus tremula*) is a medium-sized tree, seldom exceeding 18 m. (60 ft.) in height and 0·6 m. (2 ft.) in diameter. It has been planted to provide timber for the match industry. The black poplars are taller trees, up to 30 m. (100 ft.) in height and 1·25–1·5 m. (4–5 ft.) in diameter. The commonest and most important in Britain is the fast-growing hybrid black Italian poplar (*P. serotina*), a large tree with a spreading, fan-like crown and a straight cylindrical bole clear of branches for 15–18 m. (50–60 ft.) when fully grown. This and some of the other hybrid black poplars have great potential for timber production on suitable sites. The well-known Lombardy poplar (*P. pyramidalis* or *P. italica*) is grown for ornamental purposes or for shelter; the timber is of no value, being coarse and knotty.

General description. Poplar wood generally is soft and light, the density varying between 0·37 and 0·53 (23 and 33 lb./ft.^{3}) in the seasoned condition (the average for black Italian poplar is about 0·45 (28 lb.); of plain appearance, nearly white or pale reddish-brown with a fine, even texture; usually straight grained; odourless and tasteless.

Technical properties. The timber seasons well and fairly rapidly. It is one of the lighter and softer of European hardwoods though its strength properties, especially toughness, are relatively high considering its low density. It is unsuitable for bending, and is not resistant to decay. Sharp, thin-edged tools are essential to obtain a good finish as the timber is inclined to be woolly. It can be stained, painted and glued satisfactorily.

Uses. Poplar is essentially a utility timber; normally it is of no decorative value. Since it withstands rough usage without splintering, it is considered more suitable than softwood for the floors of lorries and railway wagons carrying stone, coal, etc., and for floors in warehouses and stores. It is considered to be the best timber for brake blocks on railway carriages. Poplar is used for crates and packing cases, wood-wool, toys and as an alternative to alder for hat blocks. Clean cylindrical logs are suitable for peeling into veneer for chip baskets and for plywood (poplar plywood is imported from Italy and Canada).
Aspen is the best timber for match splints and match boxes; the match industry relies mainly on supplies from Northern Europe and Canada.
Poplar wood is suitable for pulping.

Poplar

Flat-cut

Reproduced actual size

Robinia

[*Robinia pseudoacacia*]

Distribution and supplies. Robinia or false acacia, commonly called acacia, is a native of North America where it is known as black locust. It was introduced to Europe in the seventeenth century and has been extensively planted in Britain and on the Continent. Formerly it was grown under forest conditions for timber production but nowadays mainly as an ornamental tree. Grown in the open it rarely exceeds 15 or 18 m. (50 or 60 ft.) in height and 0·6 m. (2 ft.) in diameter, commonly with a twisted or fluted trunk which is inclined to fork close to the ground, so it is not usually obtainable in long lengths.

General description. The yellowish sapwood is usually less than $\frac{1}{2}$ in. wide, contrasting sharply with the heartwood which is greenish when first cut, turning golden-brown on exposure. It is usually straight grained, the texture being fairly coarse owing to the contrast between the porous springwood and the dense summer wood. These alternating bands of light and dark wood give an attractive appearance. The density is variable, usually between 0·64 and 0·80 (40 and 50 lb./ft.^{3}).

Technical properties. Robinia has valuable technical qualities. It is nearly as tough as ash, and equal to oak in other strength properties. It is excellent for steam bending and is highly resistant to decay. It seasons slowly with a marked tendency to warp. Though rather hard and heavy the timber can be worked satisfactorily by most hand and machine tools, and takes a good finish. It is not easy to nail.

Uses. In Britain supplies are small and irregular so the timber is seldom offered for sale. It is mainly used on estates for gate-posts, fencing and similar purposes. It makes a handsome furniture wood, similar to satinwood in appearance, but as logs are usually small and often rotten at the heart it is only obtainable in small dimensions and is rarely utilised in this way. In France, where the timber is more plentiful, and its outstanding strength and durability are appreciated, it is also used as a substitute for ash, for wheelwrights' work, agricultural implements, ladder rungs, etc. A traditional use is for the wooden pins (tree nails) used in shipbuilding.

Robinia

Quarter-cut

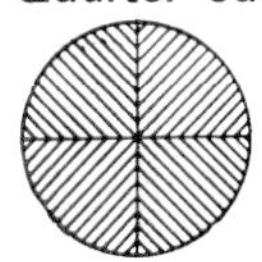

Reproduced actual size

Sycamore

[*Acer pseudoplatanus*]

The sycamore is really a species of maple (*Acer*); in some parts of Britain it is called great maple or, less appropriately, Scotch plane. It should be noted that in the USA the name sycamore refers to the American western plane (*Platanus occidentalis*).

Distribution and supplies. Sycamore is a native of Continental Europe and has become thoroughly naturalised in Britain, growing to a height of over 30 m. (100 ft.) and a diameter of 1·5 m. (5 ft.), with a straight cylindrical bole under forest conditions. Well-grown sycamore trees command a high price. Flat-sawn boards of good width and quarter-sawn stock to order are obtainable at most hardwood mills. Logs with curly or wavy grain are usually converted to veneers, either by slicing or rotary cutting.

General description. A high-quality, special-purpose timber with a fine, even texture, white or yellowish-white with a natural lustre which is especially marked when cut on the quarter. Density about 0·61 (38 lb./ft.^{3}), i.e. somewhat lighter than beech and oak. Generally straight grained but some trees develop a wavy grain which produces an attractive type of figure, known as fiddle-back, from its traditional use for the backs of violins and similar instruments. If the timber is dried slowly it assumes a light-brown colour and is then known as weathered sycamore. It is sometimes stained grey with a solution of an iron salt and marketed under the name of harewood.

Seasoning and movement. Sycamore requires care in seasoning; the surface should be dried as rapidly as possible to prevent discoloration.

Strength and bending properties. Strength properties are similar to those of oak. Straight-grained sycamore is good for steam bending but because of the prevalence of irregular grain and small knots careful selection is necessary.

Durability and preservative treatment. The timber is very susceptible to fungal decay and deteriorates rapidly if left in the log. It is not suitable for outside work unless protected from the damp, or treated with a preservative.

Working and finishing properties. When straight grained it is not difficult to work and can be planed to a fine, smooth finish. It is a very good turnery wood, takes stain, polish and other finishing treatments well, and can be glued without difficulty. It is good for carving.

Uses. The principal demand for sycamore is for various kinds of turnery, from small bobbins to the large rollers, known as bowls, used in the textile industry. Its clean, white appearance and hard, smooth surface make it specially suitable for purposes where cleanliness is imperative, such as kitchen and dairy utensils, chopping boards and bread boards. Large, clean butts are in demand for modern light-coloured furniture, especially for contract work for furnishing hotels, restaurants and ships. Figured logs are converted to decorative veneer for furniture and interior joinery. As flooring, it has a pleasing appearance and a high resistance to wear, but in the latter respect it is not equal to Canadian and Japanese maple.

Other species of interest. *The native field maple (*A. campestre*) is a common hedgerow tree, sometimes reaching commercial size. The wood is similar to sycamore. Maple from Europe (*A. platanoides*) is used as an alternative to sycamore in the manufacture of musical instruments. Rock maple or hard maple (*A. saccharum*), from Canada and the USA, is appreciably harder and heavier than sycamore; the same applies to Japanese maple.*

Sycamore

Flat-cut

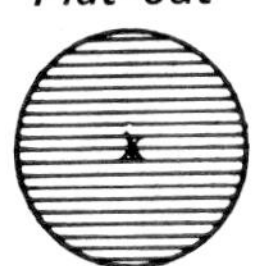

Reproduced actual size

Rotary-cut

Quarter-cut

Walnut

[*Juglans regia*]

Distribution and supplies. The European walnut, a native of Eastern Europe and South-West Asia, has been cultivated in Western Europe since early historic times. The timber is highly valued both for its ornamental and technical qualities. The demand for English walnut exceeds the supply, and much of the walnut used in Britain is imported in the form of unedged sawn timber and veneers, mainly from France, Italy, Turkey and Yugoslavia.

The tree is generally grown in the open where it may reach a height of 18–25 m. (60–80 ft.) with large spreading branches; the bole rarely exceeds 6 m. (20 ft.) in length and 1 m. (say, 3 ft.) in diameter. The timber is stocked in the form of unedged boards, 25–100 mm. (1–4 in.) thick, of varying length, mostly short. Branches with heartwood only a few inches in diameter are utilised for turnery and small parts of furniture. The entire rootstock may be excavated for veneer production.

General description. The colour of the heartwood is variable with a greyish-brown background and contrasting irregular dark streaks. On wide boards three distinct zones are seen: the true heartwood in the middle, darker than the rest; the sapwood on the outside, much paler; between them an intermediate zone, often slightly pinkish-brown in colour. When exposed to strong sunlight the colour fades, eventually becoming a dirty grey unless protected. Blue-black stains may develop on the wood if in contact with iron or steel under damp conditions. The ornamental character of the heartwood is often accentuated by a natural wavy grain. Highly figured veneers are cut from burrs, crotches and stumpwood.

French walnut tends to be somewhat paler and greyer than English, while the Italian wood is characterised by its elaborate figure and dark, streaky coloration.

Technical properties. Walnut is unique in its combination of moderate density, about 0·64 (40 lb./ft.^{3}), good working and finishing properties, toughness and resistance to splitting—qualities which make it unsurpassed for gun and rifle stocks and certain kinds of wood carving. It seasons well, though rather slowly, with relatively little shrinkage and a moderate amount of movement in service. It is classed as very good for bending. The heartwood is moderately resistant to insect and fungal attack.

Uses. Walnut is now too scarce to be used in the furniture industry in the solid form for any but high-quality cabinet work and chairs and small parts such as handles and fittings. Walnut veneers are widely used for cabinet work, panelling, shopfitting and high-class joinery. For the solid parts of veneered walnut furniture it is general practice to use other timbers such as African walnut, beech and abura, stained if necessary to match the veneer. Other uses include fancy goods, especially turnery, sports goods—mainly for decorative purposes—and musical instruments.

Other species of interest. *Black walnut (*J. nigra*) comes from the USA. Some Japanese walnut (*J. sieboldiana*) is exported to Europe.*
*There are a number of other timbers botanically unrelated to true walnut but sufficiently resembling it to be known as walnut and used for similar purposes. The most important are the so-called African walnut (*Lovoa trichilioides*) described on* *p. 80**, and Queensland walnut (*Endiandra palmerstonii*). New Guinea walnut (a species of* Dracontomelum*) and Brazilian walnut or imbuya (*Phoebe porosa*) are sometimes available. Satin walnut (*Liquidambar styraciflua*), also known as American red gum, was at one time used for bedroom furniture. East Indian walnut or kokko (*Albizia lebbek*) is rarely seen nowadays.*

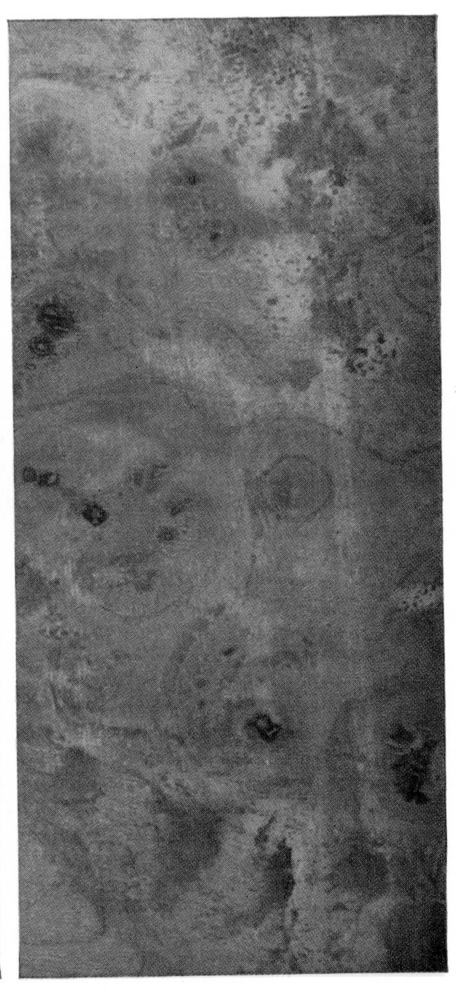

Walnut

Quarter-cut

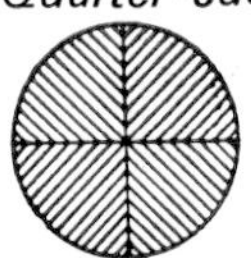

Flat-cut

Burr

Reproduced actual size

Willow

[various species of *Salix*]

Distribution and supplies. Of the numerous species and varieties of European willow, three may attain the dimensions of large trees, 21–28 m. (70–90 ft.) in height and 1–1·2 m. (3–4 ft.) in diameter. They usually grow on the banks of rivers and streams where the soil is moist and well drained. The white willow (*Salix alba*) and the crack willow (*S. fragilis*) are commonly grown as pollards, i.e., they are cut back 2·5–3 m. (8 or 10 ft.) above the ground to produce—out of reach of cattle—a growth of long straight branches, traditionally used for making fencing stakes and hurdles. The cricket-bat willow (a variety of the white willow) is the only one regularly cultivated for its timber. Under favourable conditions it grows extremely quickly, reaching the optimum size of about 0·45 m. (18 in.) diameter in 10 to 14 years. Osiers are shrubby species of willow with tough, pliant branches used for making baskets. They are grown in osier beds and are cut down to the ground every year.

General description. The timber is typically straight grained with a fine, even texture, similar to poplar in general appearance. Very light in weight; well-grown cricket-bat willow has a density of 0·34–0·42 (21–26 $lb./ft.^3$), seasoned; the other species are somewhat heavier. The nearly white sapwood is particularly wide in fast-grown cricket-bat willow and white willow, the heartwood pinkish. The annual rings are marked by fine lines, distinctly visible on the face of a cricket bat for example.

Technical properties. Seasoning presents no special difficulty. The wood is soft and not particularly strong except for its toughness; it tends to dent rather than split under rough usage (the surface of a cricket bat is compressed during manufacture to make it harder). The timber is not resistant to decay. The sapwood readily absorbs preservatives but the heartwood is resistant to treatment. Conversion and working are easy but sharp tools are essential to obtain a good smooth finish as the timber tends to be fairly woolly. It gives good results with the usual finishing treatments and can be glued satisfactorily.

Uses. The most important use for willow is in cricket bats. To be acceptable for first-quality bats the timber must be of the appropriate variety, and grown under conditions which produce white, straight-grained wood with uniformly wide annual rings, 10–20 mm. (say, $\frac{3}{8}$–$\frac{3}{4}$ in.), free from defects and blemishes (see the Forest Products Research Laboratory report *Timbers used in the Sports Goods Industry*). Timber which does not meet this strict specification may be used for second-quality and toy bats, artificial limbs, the floors of lorries and railway wagons, brake blocks, clog soles, sieve rims, medicinal charcoal, chip baskets and wood pulp.

Willow

Flat-cut

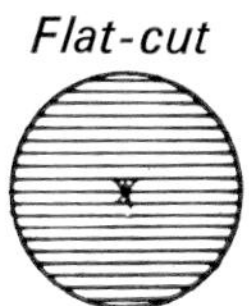

Reproduced two-thirds actual size

European Timbers

SOFTWOODS

Cedar

[three species of *Cedrus*]

The name cedar is applied to several different kinds of timber, both hardwood and softwood, characterised by a natural fragrance. Cedar of Lebanon (*Cedrus libani*) and the closely allied species Atlantic or Atlas cedar (*C. atlantica*), of Algeria and Morocco, and the Himalayan deodar (*C. deodara*) are regarded as the true cedars.

Distribution and supplies. The cedars can scarcely be classed as commercial timbers outside their countries of origin, but they are of sufficient interest to be included, being commonly planted for ornamental purposes in the grounds of large houses where they grow to an enormous size, 30 m. (100 ft.) in height and 2·5 m. (8 ft.) in diameter at the butt, with spreading branches. The timber is not regularly marketed but can often be obtained from timber merchants.

General description. The heartwood is strongly scented when freshly cut, and somewhat resinous. It is light-brown, with a pleasing appearance due to the irregular pattern of the annual rings. Seasoned timber has a density of about 0·56 (35 lb./ft.^{3}).

Technical properties. The outstanding technical characteristic of cedar is its unusually high resistance to decay and insect attack. It is not particularly strong and is inclined to be brittle. As the trees are almost invariably grown in the open, with heavy branches, a fairly large proportion of the converted timber is coarse and knotty, with irregular grain. Clear material works easily by hand and machine tools, and takes a good finish.

Uses. The timber is eminently suitable for outdoor use in the form of garden furniture, gates and fences and other purposes where resistance to decay is the main requirement. Good-quality material is used for clothes chests and cupboards, on account of the pleasant smell, and for interior joinery and decorative veneer.

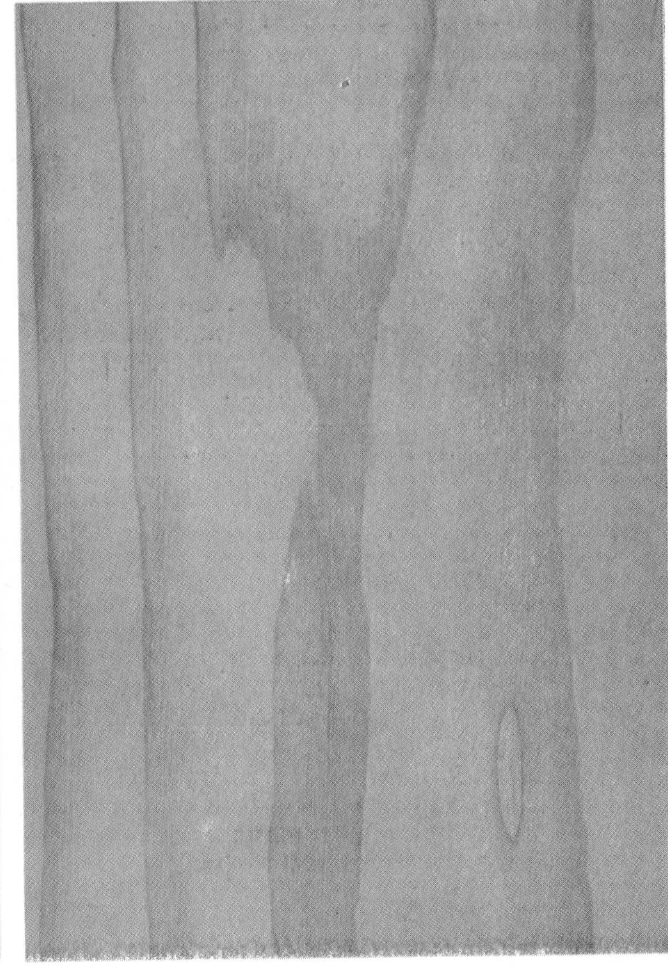

Cedar

Quarter-cut

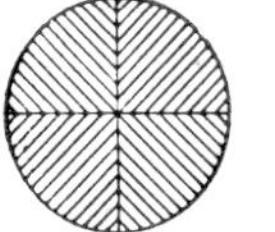

Reproduced actual size

Flat-cut

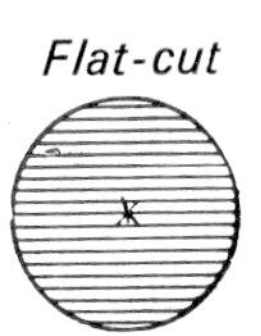

Larch

[*Larix decidua*]

Distribution and supplies. Although a native of the Alps, the European larch has long been a major forest species in Britain also. When full grown it is a large tree, up to 40 m. (130 ft.) in height and 1 m. (say, 3–4 ft.) in diameter, but rarely exceeds 30 m. (100 ft.) in plantation.

General description. Larch is distinguished from other softwoods in common use by the hard, dense character of the reddish-brown resinous heartwood, which is sharply differentiated from the narrow sapwood. The annual rings are clearly marked by the dark summer wood zones. Except for pitch pine, larch is the heaviest of the softwoods in common use. The average density of mature timber after seasoning is about 0·69 (37 lb./ft.^{3}). Wood from young trees of vigorous growth is considerably lighter, 0·51–0·55 (say, 32–34 lb./ft.^{3}).

Seasoning and movement. For a softwood, larch dries fairly slowly and tends to twist, check and split. Once it is thoroughly seasoned the movement in service is small.

Strength and bending properties. Larch is one of the strongest softwoods in common use, being considerably harder than Baltic redwood, and slightly stronger in bending strength and toughness.

Durability and preservative treatment. Larch also ranks high in its resistance to decay, and has the advantage of a comparatively small proportion of perishable sapwood. It lasts longer than Scots pine or Baltic redwood but not so long as oak. Where a long life is required under exposed conditions, e.g., as fencing, preservative treatment is well worth while, particularly if the timber is to be used in the round.

Working and finishing properties. Larch is more difficult to saw and work than most softwoods. Clear, straight-grained material finishes cleanly but the hard knots are liable to damage the edges of cutting tools, and the surface tends to tear where the grain is irregular. The wood requires care in nailing, to avoid splitting.

Uses. Because of its superior strength and lasting qualities, larch is particularly useful for outdoor work or where a timber stronger and more durable than ordinary deal is required. Larch is grown mainly on private estates for use in the construction of farm buildings, gates and fences. With the increasing use of wood preservatives, natural durability is less important than it used to be; nevertheless the advantage of larch in this respect is still appreciated where treated timber is not readily available. In building larch, is used mainly for exterior work such as cladding, gates and sills. It is a traditional timber for boat building, particularly in Scotland, and is also used for the floors of motor lorries and for vats in the chemical industry. A great deal of larch is used in the form of poles. Thinnings of all sizes can be utilised profitably for pergolas and rustic work in gardens, and as stakes, bean rods, etc.

Other species of interest. *In recent years Japanese larch (*L. leptolepis*) has been widely planted in Britain, in preference to the European species. Comparative tests on European and Japanese larch of similar age, however, have indicated that, age for age, there is probably little to choose between them. Siberian larch (*L. sibirica*) has been exported in small quantities from Kara Sea ports and the Archangel district. The timber is very narrow ringed, of joinery quality; it has been used for the construction of vats and for joinery.*

Larch

Flat-cut

Reproduced actual size

Redwood or Scots Pine

[*Pinus sylvestris*]

Timber of this species imported into Britain from the Continent is commonly called redwood (not to be confused with California redwood, *Sequoia sempervirens*), alternatively red or yellow deal or fir, with qualifying names indicating the geographical origin, such as Baltic, Finnish, Swedish, Archangel, White Sea, etc. Timber grown in Britain is known as Scots pine, Scotch fir, fir, or home-grown deal.

Distribution and supplies. The tree is widely distributed on the Continent and has been planted extensively in Britain for timber production. Under favourable conditions it reaches 30 m. (100 ft.) in height and 0·6–0·9 m. (2–3 ft.) in diameter but most of the commercial timber from the forests of Northern Europe is cut from very much smaller trees. The timber is imported in large quantities and a wide range of sizes, as deals, battens, boards, etc., railway sleepers, poles and pit props. Widths are never more than 280 mm. (11 in.), and the prices of 180 mm. (7 in.) and wider material are very much higher than those for 150 mm. (6 in.) and narrower. Lengths in scantling sizes commonly run up to 5 m. (17 ft.).

General description. The wide geographical range of this species is reflected in the variable character of the timber, particularly the rate of growth (ring-width), the texture of the wood and the number and size of knots. In the seasoned condition the lighter-coloured sapwood is usually distinct from the pale reddish-brown resinous heartwood; the sapwood comprises a relatively large proportion of the converted timber. The annual rings are clearly marked by the contrast between the light spring wood and the darker summer wood zones. The average density of seasoned timber is about 0·48 (30 lb./ft.^{3}). The wood is usually straight grained except in the neighbourhood of knots, which occur at fairly regular intervals.

Seasoning and movement. The timber can be dried rapidly without serious degrade. To reduce the risk of blue stain in the sapwood, the surface of converted timber should be dried quickly; alternatively, it may be dipped in an antiseptic solution as it comes off the saw. The movement is classed as medium.

Strength properties. The strength depends largely on the incidence of knots and other natural defects which are taken into account in grading timber for constructional work. Permissible working stresses for graded timber in structural sizes are given in the British Standard Code of Practice, No. 112 *The Structural Use of Timber in Buildings.* Redwood is slightly stronger than whitewood (spruce) of similar grade and less strong than Douglas fir, pitch pine and larch.

Durability and preservative treatment. Redwood is classed as non-durable. The heartwood is moderately resistant to preservative treatment; the sapwood is permeable. Untreated sapwood is susceptible to attack by the common furniture beetle.

Working and finishing properties. In general the timber works easily in all hand and machine operations. Clear, narrow-ringed material showing more than 5 rings to the centimetre (say, 13 rings to the inch), as used for joinery, takes a good clean finish from the tool. Where the rings are appreciably wider, however, there is a tendency for the soft spring wood to tear up, especially if the cutters are dull. Redwood takes nails well and, with the exception of the occasional resinous piece, can be stained and glued effectively and gives good results with paint, varnish and polish.

Uses. Being comparatively cheap, easily worked and available in a wide range of sizes, redwood is the principal material used in Northern Europe for carpentry and joinery, boxes and packing cases and many other general purposes. In the round it is the standard timber for transmission poles and pit props. It is pulped for kraft paper.

Redwood

Flat-cut

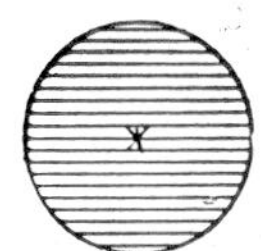

Reproduced actual size

Quarter-cut

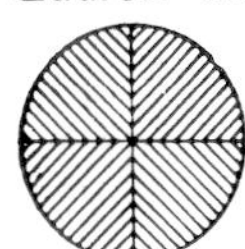

Whitewood or Spruce

[*Picea abies*]

In Britain imported timber of European or Norway spruce goes by the name of whitewood, white deal or white fir, sometimes qualified to indicate the country of origin or port of shipment, for example Baltic, Finnish, Swedish, Archangel, Yugoslavian, etc.

Distribution and supplies. The species is distributed throughout Northern Europe, and has been widely planted for timber production in Britain. It is familiar to everyone as the Christmas tree. It is a tall tree exceeding 45 m. (150 ft.) in height and 1·25–1·5 m. (4–5 ft.) in diameter under favourable conditions. The timber is sawn and graded for export from the principal producing countries in much the same way as redwood and can generally be supplied to the same specifications.

General description. As the name implies, whitewood is nearly white, with no visible difference between sapwood and heartwood. The annual rings are less conspicuous than in redwood, and the timber is less dense, average about 0·42 (26 lb./ft.^{3}). It is usually less resinous than redwood. There is a tendency to spiral grain, especially in timber cut from small trees. In contrast to redwood, the knots are irregularly distributed. The British-grown timber is essentially similar to the imported but much of it is wide-ringed, coarse and knotty.

Seasoning and movement. The timber seasons rapidly and well, with little tendency to split or check. Spiral-grained material from small logs is inclined to twist in drying. The movement of whitewood under varying conditions of humidity is classed as small, i.e., it is somewhat more stable than redwood.

Strength properties. Grade for grade, the imported timber is only slightly inferior to redwood in strength, and is placed in the same group in the British Standard Code of Practice, No. 112 *The Structural Use of Timber in Buildings.* The British-grown timber falls in a lower strength class; it should therefore be used in larger dimensions where strength is important.

Durability and preservative treatment. Whitewood is inferior to redwood in its resistance to decay; moreover it is not easily treated with preservatives even under pressure. Untreated sapwood is susceptible to attack by the furniture beetle.

Working and finishing properties. In general the timber works very easily with all hand and machine tools. Clear, narrow-ringed material of joinery grade finishes well provided that the cutters are sharp; wide-ringed wood is inclined to tear, and the hard dead knots are liable to damage the cutting edges of tools. The timber takes nails very well and gives good results with stain, paint and glue.

Uses. Whitewood is used for many of the same purposes as redwood. In Scotland and the North of England it is largely used for structural work in housebuilding. Being more susceptible to decay and resistant to preservative treatment it is not so suitable for outside work. Because of its clean, white appearance and lack of odour it is often preferred to redwood for interior joinery, boxes and packing cases, food containers and wood-wool. Half-round ladder sides are usually made from spruce poles. Spruce is the principal European species used for pulp.

Other species of interest. *Sitka spruce (*Picea sitchensis*) has been widely planted in Britain; it is used in much the same way as British-grown Norway spruce. The Sitka spruce imported from British Columbia is a different commodity, a large proportion of the timber being of joinery quality. Eastern Canadian spruce is comparable to Baltic whitewood.*

Whitewood

Flat-cut

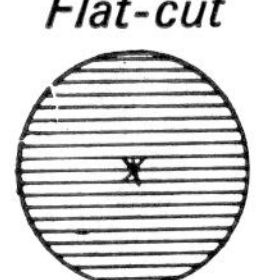

Reproduced actual size

Yew

[*Taxus baccata*]

Distribution and supplies. The common yew is widely distributed in the north temperate regions of Europe and Asia. In the British Isles it is quite commonly found growing wild, especially on chalky soils. From time immemorial it has been planted in churchyards and, more recently, in gardens. In spite of the technical and aesthetic value of the wood it is seldom planted for timber production; the reasons are the poor form of the tree and its slow rate of growth. In Britain it is generally a small, densely branched tree rarely exceeding 12 m. (40 ft.) in height. The trunk of old trees is usually more or less fluted, and often consists of several vertical shoots which have grown together, enveloping the original main stem. The yew attains a great age; trees 1·5 m. (5 ft.) or more in diameter are not uncommon. Local supplies are usually sufficient to meet the limited demand.

General description. The narrow sapwood is almost white, in sharp contrast to the heartwood, which is bright orange-brown or purplish-brown when freshly cut, toning down to a warm brown shade. It is often marked with dark streaks and small black knots. For a softwood it has a remarkably fine, even texture. The annual rings are usually narrow and inclined to be irregular, giving the wood its characteristic decorative figure.

Technical properties. Though technically a softwood (conifer), yew has many of the qualities of a hardwood. It is hard and heavy, average density about 0·67 (42 lb./ft.^{3}), seasoned, and both tough and resilient, approaching oak in these respects. It is one of the best softwoods for steam bending. Yew is very resistant to decay and is probably the most durable British timber, being practically immune to fungal attack, though damage by the common furniture beetle has been recorded. It seasons well with little distortion. When straight grained it is not difficult to work, having regard to its density, and is particularly suitable for turning, taking a clean, glossy finish and a high polish.

Uses. The principal uses of the timber depend on its outstanding strength, durability and handsome appearance. It has the disadvantages of being wasteful in conversion and relatively hard to work, and owing to the form of the tree it is not readily obtainable in large sizes. Yew is the traditional timber for archery bows and is still used for this purpose, though it has been largely superseded by degame from the West Indies and Central America. Its lasting qualities in the form of gate- and fence-posts are proverbial. Yew makes beautiful furniture and was formerly in regular use for chairs, especially Windsor chairs, and for small articles of furniture. Nowadays it is used to a limited extent in the solid and as veneer, for high-quality furniture and cabinet work, especially reproductions, and for fancy turnery.

Yew

Flat-cut

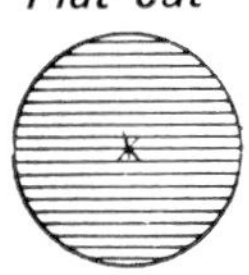

Reproduced two-thirds actual size

African Timbers
HARDWOODS

Abura

[*Mitragyna ciliata*]

The Nigerian name abura has been adopted as the British Standard name for the commercial timber; it is also known as bahia (France and the Ivory Coast) and subaha (Ghana).

Distribution and supplies. Abura is widely distributed in West Africa. It is common in the freshwater swamp forests of the evergreen forest zone (e.g., in Southern Nigeria), so supplies are plentiful and relatively cheap, competing with beech in price. By West African standards the tree is of only moderate size. Export logs are usually 0·4–0·75 m. (16–30 in.) in diameter and 3·5–5·5 m. (12–18 ft.) in length. Square-edged sawn timber is 150 mm. (6 in.) and up wide (average 200–230 mm. or 8–9 in.) and 2 m. ($6\frac{1}{2}$ ft.) and up long (average about 2·5 m. or 8 ft.).

General description. A light hardwood of plain appearance. The wide sapwood is not usually differentiated from the light yellowish-brown or pinkish-brown heartwood but some logs show a small, irregular greyish-brown heart which in square-edged stock may give the appearance of an edging of stained sapwood. The grain is inclined to be irregular. The texture is fine and uniform with a dull, lustreless surface. The average density is about 0·56 (35 $lb./ft.^3$) in the seasoned condition.

Seasoning and movement. It seasons rapidly, with little warping or splitting, and is stable under varying conditions of atmospheric humidity.

Strength and bending properties. In strength properties abura is similar to African mahogany. It is not suitable for steam bending.

Durability and preservative treatment. Logs are liable to be attacked by pinhole borers. The timber is not durable; the heartwood is rather resistant to impregnation but it is not normally used under conditions that call for preservative treatment.

Working and finishing properties. Abura is reputed to be one of the best West African timbers for machining and the manufacture of small mouldings. It works well and cleanly with hand and machine tools if the cutters are sharp and machines are in good condition. Its silica content, however, sometimes tends to blunt the cutters. The timber takes a satisfactory finish, can be stained, painted (using normal primers) or polished, and is readily cut into veneer. It holds nails and screws well and glues satisfactorily.

Uses. This timber is useful for light constructional work where no great strength is required. It is used for the interior parts of furniture (drawer sides, runners and framing) and for matching with more decorative timbers, especially walnut; also for mouldings, interior joinery and fittings, and domestic woodware. It is suitable for flooring where pedestrian traffic is light.

Abura

Flat-cut

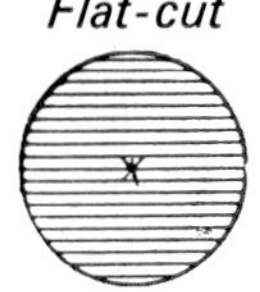

Reproduced actual size

Afara or Limba

[*Terminalia superba*]

Afara is the common Nigerian name and a British Standard name for the commercial timber. The alternative standard name limba is generally used for timber originating in the Congo or in those territories formerly under French protection, which are the main sources of supply. The plain light-coloured wood, as illustrated, is called light afara, light limba, limba clair or limba blanc; the figured heartwood, dark afara, dark limba, limba noir or limba bariolé. Other trade names are ofram (Ghana), akom (Cameroon), fraké (Ivory Coast), noyer du Mayombe (France) and korina (a registered name in the USA).

Distribution and supplies. This species is widely distributed in the West African forest zone, and is one of the more common trees of the region. It is a tall tree with a long, straight, cylindrical stem, 0·9–1·5 m. (3–5 ft.) in diameter above the buttresses. Supplies are reported to be nearly unlimited. It is shipped in the form of logs 0·6–0·9 m. (2–3 ft.) in diameter and 3·5–7·5 m. (12–24 ft.) long, mainly for plywood manufacture, and as square-edged sawn timber in a wide range of sizes, 25–50 mm. (1–2 in.) thick, 150–250 mm. (6–10 in.) wide, in lengths of 2–5 m. ($6\frac{1}{2}$–$16\frac{1}{2}$ ft.).

General description. The wood is a uniform light yellowish-brown throughout, resembling light oak in colour and mahogany in grain, but some logs have a core of coloured wood—sometimes grey but often quite dark, sometimes uniform in colour but commonly with darker banding, and, occasionally, highly figured, suggesting walnut. The density is variable, usually between 0·48 and 0·64 (30 and 40 lb./ft.^{3}), seasoned, but occasionally heavier; when lighter in weight it is commonly affected by brittleheart.

Seasoning and movement. The timber dries quickly and with little degrade; once dry, it is rated as having a small movement in service.

Strength properties. The strength is rather variable, depending on the density. It is extremely prone to brittleheart and should not be used for structural purposes unless carefully inspected for this defect.

Durability and preservative treatment. Afara is a non-durable timber, susceptible to both insect and fungal attack. Logs are liable to be attacked by pinhole borer beetles immediately the tree is felled, so special precautions are necessary to prevent deterioration.

Working and finishing properties. A mild wood which presents little difficulty in working, whether with machine or hand tools. In planing, it is an advantage to employ a cutting angle of 20°, to reduce the tendency to tear when irregular grain is present, but otherwise a good finish is obtained in most operations if care is taken. There is a slight tendency to split in nailing and screwing. The wood takes stain readily and can be glued satisfactorily. It paints well with normal primers, previously thinned.

Uses. It is mainly used for general-purpose plywood. Selected figured material is sliced for decorative veneers. The figured wood has never been popular in Britain but the plain light-coloured timber is used to a limited extent in furniture production —mainly for framing, lipping and drawer sides, and for interior joinery such as library shelves and shop fitting. It is more popular on the Continent. Where a good appearance is required, worm-free timber should be specified. Wormy grades can be used for core stock.

Afara/Limba

Quarter-cut

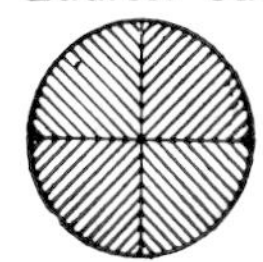

Reproduced actual size

'African Walnut'

[*Lovoa trichilioides* formerly called *L. klaineana*]

Botanically *Lovoa* is included in the mahogany family. Since it has been found suitable as a substitute for walnut it has been generally known in Britain as 'African walnut'. The French timber trade has adopted the more distinctive name of dibétou, alternatively noyer d'Afrique. Trade names current in the USA include tigerwood, congowood, alona wood and lovoa wood.

Distribution and supplies. The tree is widely distributed in the mahogany districts of West Africa—from Sierra Leone to Gaboon—and regularly shipped with consignments of mahogany from the principal ports. It is a large tree with a well-shaped, cylindrical bole providing logs up to about 1 m. (say, 40 in.) in diameter, 3·5–7 m. (12–23 ft.) or more in length. Sawn timber is produced in a range of sizes, up to 350 mm. (14 in.) wide (square-edged) and 3·75 m. (12 ft.) long. It is also available in veneer form.

General description. In grain and texture the timber resembles African mahogany. It is readily distinguished by its yellowish-brown or golden-brown colour, as compared with the reddish- or pinkish-brown of mahogany, and is sometimes marked with dark streaks or veins; these markings are said to be more prominent in logs of irregular shape. The grain is usually more or less interlocked, and when cut on the quarter this results in a distinct ribbon or stripe figure as in African mahogany. The resemblance to walnut is closer in flat-sawn timber. The average density is about 0·55 (34 lb./ft^3.), seasoned.

Seasoning and movement. The timber can be dried fairly rapidly, without much degrade, though existing heart shakes may extend. It is a stable timber; movement in service is classed as small.

Strength and bending properties. The results of limited tests indicate that the timber is superior to African mahogany and inferior to true walnut in strength and bending properties.

Durability and preservative treatment. On arrival logs are sometimes damaged by pinhole borers and longhorn beetles. The timber has been found to be moderately resistant to fungal decay in Britain, and moderately resistant to termites in Nigeria. The heartwood is extremely resistant to preservative treatment.

Working and finishing properties. The timber is not difficult to work although quarter-cut material with interlocked grain has a tendency to pick up in machining; otherwise it finishes cleanly if tools are kept sharp. By comparison with American black walnut it is about 20 per cent easier to cut and has 50 per cent less dulling effect on cutters. It takes nails and screws well and can be stained to match with either walnut or mahogany. Gluing is satisfactory. The wood is not usually painted but, if needed, a thin normal primer or aluminium sealer is recommended.

Uses. African walnut is mainly used as a substitute for true walnut for furniture and high-class joinery, especially for the solid parts of furniture veneered with European walnut.

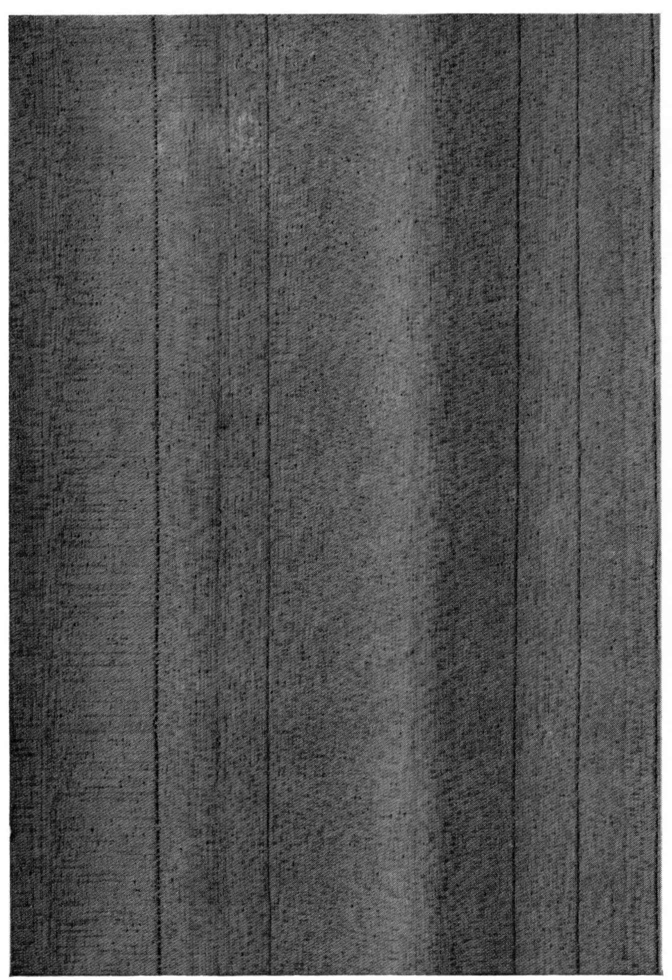

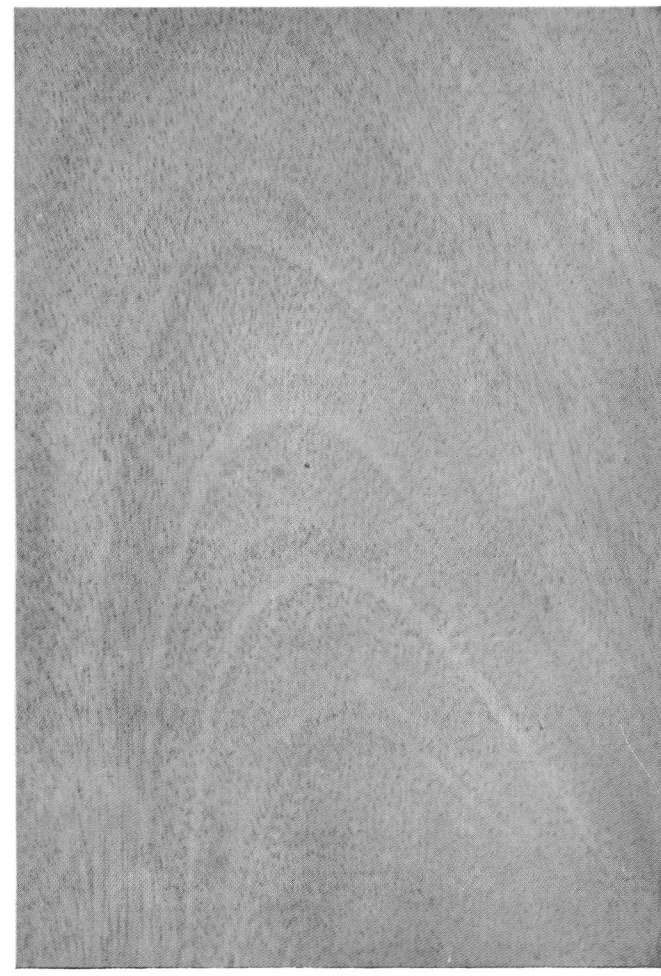

African Walnut

Quarter-cut

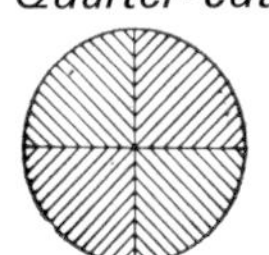

Reproduced actual size

Flat-cut

Afromosia

[*Afrormosia elata*]

This timber was introduced to the timber trade from Ghana after the second world war under the local vernacular name of kokrodua. It is now generally known as afrormosia.

Distribution and supplies. The tree has a limited distribution in the dry forest zone of Ghana, the Ivory Coast, Nigeria and the Congo. The timber is available in log form, up to 1·5 m. (5 ft.) diameter and 10 m. (33 ft.) long, as square-edged sawn material, 150–450 mm. (6–18 in.) wide and up to 6 m. (20 ft.) long, and as decorative veneer. In response to popular demand, exports, mainly from the Ivory Coast and Ghana, have been greatly increased in recent years to surpass those of iroko.

General description. A high-quality timber with an attractive appearance, resembling a fine-grained teak but without the latter's oily nature. The heartwood is yellowish-brown, somewhat variable in shade as between one log and another, creating difficulties in matching. Unlike teak, it does not bleach on exposure to the weather but tends to darken with time. It is liable to stain if in contact with iron under damp conditions. The grain varies from straight to interlocked. The density in the seasoned condition averages about 0·70 (44 $lb./ft.^3$), i.e., somewhat heavier than teak.

Seasoning and movement. Afrormosia seasons fairly slowly but well, with little degrade, and keeps its shape extremely well, though less so than teak.

Strength and bending properties. It is appreciably stronger and harder than the general run of teak. A tough, shock-resistant wood, superior to beech but not quite as good as ash or rock elm in this respect. For steam bending it is classed as only moderate.

Durability. Afrormosia is resistant to fungi, insects and marine borers and is placed in the top durability class with teak, iroko and other high-ranking timbers.

Working and finishing properties. In spite of its hardness, the timber can be worked fairly readily with hand or machine tools, and the blunting effect on cutters is small in comparison with teak. Where the grain is interlocked, machined surfaces are liable to tear; otherwise the wood finishes cleanly and gives good results with stains, polishes and glue. It tends to split when nailed.

Uses. Afrormosia is recognised as an alternative to teak for ship- and boat-building and other purposes requiring high strength properties, stability, durability and a good appearance. It is widely used for high-class furniture, cabinet work and joinery, in the solid form and as veneer, the one disadvantage being the variation in colour between one log and another. To prevent the staining liable to occur when the wood is in contact with iron or steel, the use of non-ferrous metal fastenings and fittings, or alternatively galvanised iron or steel, is recommended.

Afrormosia

Quartercut

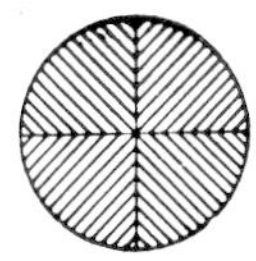

Flat-cut

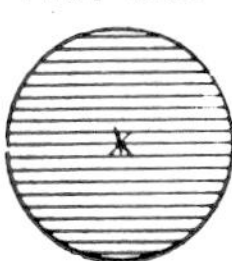

Reproduced actual size

Afzelia or Doussié

[various species of *Afzelia*]

The genus *Afzelia* is represented by five species in West and Central Africa. Timber shipments are believed to be mainly *A. bipindensis* and *A. pachyloba*; varying proportions of *A. africana* may be included. The East African species is *A. quanzensis*. Afzelia is the recommended British Standard name for the timber of all species. In practice, however, it is generally marketed under the names current in the countries of origin: doussié (Cameroon), apa (Nigeria), chamfuta or mussacossa (Mozambique), mkora or mbemba-kofi (Tanzania).

Distribution and supplies. The genus *Afzelia* is widely distributed in tropical Africa. The largest trees are found in the evergreen forests of West and Central Africa; the timber is shipped from West African ports (mainly in Cameroon and Nigeria) as logs 0·6-1 m. (24–40 in.) in diameter and 3·5–7·5 m. (say, 12–24 ft.) long, and is also available as square-edged sawn material 25–75 mm. (1–3 in.) thick, up to 300 mm. (12 in.) wide and 4 m. (13 ft.) long. The Cameroon timber is reputed to be the best. It has also been exported in small quantities from Mozambique and Tanzania.

General description. The timber has a handsome, if somewhat plain appearance. It has some resemblance to iroko in grain and texture, but is reddish-brown in colour, and with an average density of about 0·82 (51 lb./ft.^{3}) when dry, it is about 25 per cent heavier than iroko. The grain is generally straight but sometimes interlocked or irregular. Deposits are a common feature. They may be white and are occasionally accumulated in hard, stony masses, giving trouble in sawing; in other logs the deposits are soft and yellow and have little effect on the technical behaviour of the timber.

Seasoning and movement. The timber can be kiln dried satisfactorily, if extremely slowly. It shrinks very little in drying, and once dry has a very small movement in service.

Strength and bending properties. A heavy wood, about 20 per cent denser than oak, afzelia is superior to oak in all strength properties except resistance to suddenly applied loads. For steam bending it is classed as only moderately good.

Durability. A very durable timber, noted for its resistance to attack by fungi, termites and marine borers.

Working and finishing properties. Afzelia presents some difficulty in working, with a resistance generally comparable to tough oak; care must be taken to avoid occasional stony deposits. It has a moderate blunting effect on cutting edges, which must be kept sharp if a satisfactory finish is to be obtained, and when finishing quartered stock with an interlocked grain a reduction in cutting angle from 30° to 20° is advisable. If filled and polished the wood takes an extremely fine finish. It is not easy to glue.

Afzelia can be sliced to produce a decorative veneer.

Uses. Being a heavy wood with excellent strength properties combined with outstanding durability and stability and a good appearance, afzelia is used for high-class joinery, both indoors and outdoors, including staircases, panelled doors, counter tops, laboratory benches, fittings in banks and public buildings, thresholds, sills and large window units. It makes an attractive floor suitable for normal pedestrian traffic. The timber is accepted as an alternative to teak where its greater weight is not an objection.

Being dense and relatively impervious it is one of the best hardwoods for vats for the chemical industry. Because of a yellow dyestuff in the wood it should not be used for draining-boards, laundry equipment or garden seats.

Afzelia

Quarter-cut

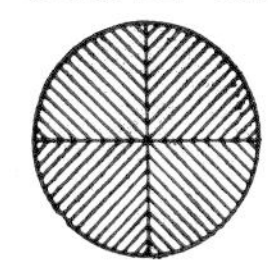

Reproduced actual size

Agba

[*Gossweilerodendron balsamiferum*]

Agba is the Nigerian name and the recommended British Standard name. Timber from the Congo may be marketed as tola and from the Portuguese Congo as tola branca or white tola. Note that tola is one of a number of names used also for the timber of *Oxystigma oxyphyllum* (see page 170).

Distribution and supplies. Agba occurs in tropical Africa from Nigeria southwards to the Congo basin. Commercial supplies are obtained mainly from Nigeria and the Congo, especially the Portuguese territory of Cabinda. The tree is one of the largest in tropical Africa, up to 60 m. (say 200 ft.) high and 1·5–2·0 m. (say 5–7 ft.) in diameter; unbuttressed, it has a straight, cylindrical bole of 24–30 m. (80–100 ft.). It is available in the form of logs up to about 1·8 m. (6 ft.) diameter, more commonly 0·5–1·0 m. (20–40 in.) and 7·5 m. (24 ft.) long, and as square-edged and unedged sawn timber 16–100 mm. ($\frac{5}{8}$–4 in.) thick, up to 450 mm. (18 in.) wide, average 230 mm. (9 in.), and 5·5 m. (18 ft.) long (average 2·75 m. or 9 ft.). Agba is also available in the form of veneer for decorative work and as plywood.

General description. Agba is one of the lighter-weight tropical African timbers, the density averaging about 0·51 (32 lb./ft.^{3}), seasoned. The pale, straw-coloured heartwood has an attractive appearance with a moderately fine and firm texture and a characteristically interlocked grain producing a fairly close stripe on quartered surfaces; the sapwood is fairly well defined, 100 mm. (4 in.) or more in width, but is often removed from logs before shipment.

The timber has a slight odour, due to a natural gum which, in occasional logs, is present in considerable quantities; the gum may cause inconvenience in sawing and handling but in general is not regarded as a serious defect. Tension wood may be present in misshapen logs, and a central core of brash timber (brittleheart) is often present in large logs.

Seasoning and movement. The timber seasons fairly rapidly, with little tendency to distort or split but with a risk of some gum exudation. Once dry, it is rated as having a small movement in service.

Strength and bending properties. Sound agba heartwood is comparable in strength to African mahogany. The timber is only moderately good for bending.

Durability and preservative treatment. Logs are sometimes damaged by pinhole borers and longhorn beetles, and the sapwood is susceptible to attack by powder-post beetles. The heartwood has a considerable measure of natural resistance to fungal attack and damage by subterranean termites; it is not readily treated with preservatives.

Working and finishing properties. Agba presents little difficulty in working, apart from some gum accumulation when sawing green or air-dry stock. A cutting angle of 20—25° is advisable when planing or moulding quartered stock, and sharp cutters are essential to ensure a clean finish. For painting, aluminium primer is recommended. Gummy material should be avoided if polish is to be applied. Gluing is reported to be satisfactory.

Uses. Agba is used for furniture, especially in schools and churches, and for interior and exterior joinery. For flooring it is suitable for light pedestrian traffic, and for buildings with floor panel heating. In boat building it is employed for interior fittings, for laminated frames and planking and, in the form of veneer, for the construction of hot-moulded hulls. For outside work with a risk of decay it is essential to eliminate sapwood, and for structural work brittleheart should be excluded. It is not recommended where the slight odour might cause taint.

Agba

Quarter-cut

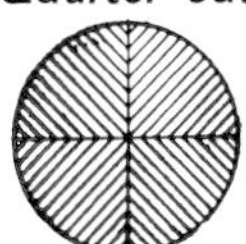

Reproduced actual size

Antiaris

[*Antiaris africana* and *A. welwitschii*]

Antiaris is the British Standard name for this timber. It is also marketed as oro (Nigeria), chen chen or kyen kyen (Ghana) and ako (Ivory Coast).

Distribution and supplies. The two species which furnish the commercial timber are widely distributed in tropical Africa, and locally abundant. They grow to a large size with a straight cylindrical stem up to 21 m. (70 ft.) long and 0·6–1·5 m. (2–5 ft.) in diameter. The timber has been shipped in small quantities from Ghana, the Ivory Coast and Nigeria, in log form and as sawn timber.

General description. In general appearance the wood has a close resemblance to obeche but is not so lustrous: it is a pale straw colour, with an interlocked grain producing a prominent figure on quartered surfaces, which gives it a distinctly decorative appearance suggesting white sapele. There is no colour contrast between the heartwood and the wide sapwood which is very susceptible to sap stain and pinhole borer damage when green, and liable to attack by powder-post beetles when dry. The wood is somewhat variable in weight, on the average about 10 per cent heavier than obeche, but may be lighter when cut from near the centre of large trees, which are sometimes affected by brittleheart or spongy heart. When green the timber has an unpleasant smell but, except occasionally in enclosed conditions, this is not troublesome when the wood is dry.

Seasoning and movement. Antiaris seasons fairly rapidly, with a strong tendency to distort. It has a comparatively small shrinkage in drying from green to 12 per cent moisture content and its movement in service is rated as small.

Strength and bending properties. Antiaris is unlikely to be used for its strength properties, which are generally similar to those of obeche, except that the latter has greater shock resistance. Limited tests indicate that the timber is unsuitable for steam bending.

Durability and preservative treatment. Rapid extraction and the use of chemical sprays or dips are essential to prevent fungal discoloration and insect attack. The seasoned timber is perishable and should not be exposed to conditions favourable to decay without preservative treatment. It is permeable to liquids, and readily treated.

Working and finishing properties. Antiaris works easily and finishes cleanly although, as for most light-weight woods, cutters must be kept sharp. In planing, tearing can be avoided if the cutting angle is reduced to 20° but there is some tendency for the grain to rise. The timber nails well, takes glue and can be stained and polished satisfactorily. It can be peeled without difficulty.

Uses. Being similar in character to obeche, antiaris is suitable for the same purposes provided it can be supplied in good condition, free from stain, insect damage and brittleheart. Uses include furniture interiors and light constructional work such as boxes, where no great strength is required; plywood and sliced veneer for decorative purposes can also be obtained.

Antiaris

Quarter-cut

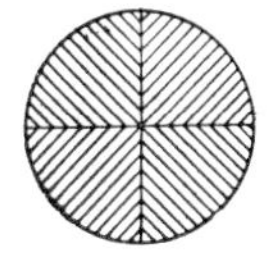

Reproduced actual size

Avodiré

[*Turraeanthus africanus*]

Botanically *Turraeanthus* is included in the mahogany family. The timber was introduced to the trade from the Ivory Coast by way of France and is known by the French name avodiré.

Distribution and supplies. Avodiré has a restricted range in West Africa. Supplies are obtained from the Ivory Coast and Ghana. It is a relatively small tree, rarely exceeding 0·6 m. (2 ft.) in diameter, of rather poor shape, and is not, therefore, ideal for lumber production. It is exported in relatively small quantities, mainly in log form or as curls for the manufacture of decorative veneer.

General description. Avodiré differs from most timbers of the mahogany family in being nearly white or pale-yellow when freshly cut, darkening to a golden-yellow shade with a natural lustre. The grain is sometimes straight but often wavy, producing an unusual mottled figure when sliced or cut on the quarter, something like Ceylon satinwood but not so fine textured. The figure of avodiré curls is particularly attractive. The plain, unfigured wood may be compared to a white mahogany or a dense grade of obeche. It is fairly soft and light in weight, resembling African mahogany in these respects; average density about 0·55 (34 lb./ft.^{3}) in the seasoned condition.

Seasoning and movement. The sawn timber can be dried fairly rapidly, with some tendency to cup and twist; shakes are liable to extend. Like most timbers of the mahogany family its dimensional movement in service is small.

Strength and bending properties. Straight-grained timber has been found to possess high strength properties, for its weight, equal to oak except in resistance to shock. Bending properties are rather variable; it is classed as very poor in this respect.

Durability and preservative treatment. Avodiré is not a durable timber and is unsuitable for use in situations favouring decay or insect attack. Moreover, the heartwood is extremely resistant to preservative treatment.

Working and finishing properties. The timber is fairly easy to work with hand or machine tools, provided that cutting edges are kept sharp. Where there is a tendency for the surface to tear or pick up due to irregular grain the cutting angle should be reduced to 20° or less. Avodiré responds well to conventional finishing treatments and can be glued satisfactorily.

Uses. The wood is probably best known as a decorative veneer for furniture and panelling. Non-figured logs are sometimes used for plywood. The sawn timber is useful for interior joinery and cabinet work, as an alternative to sycamore.

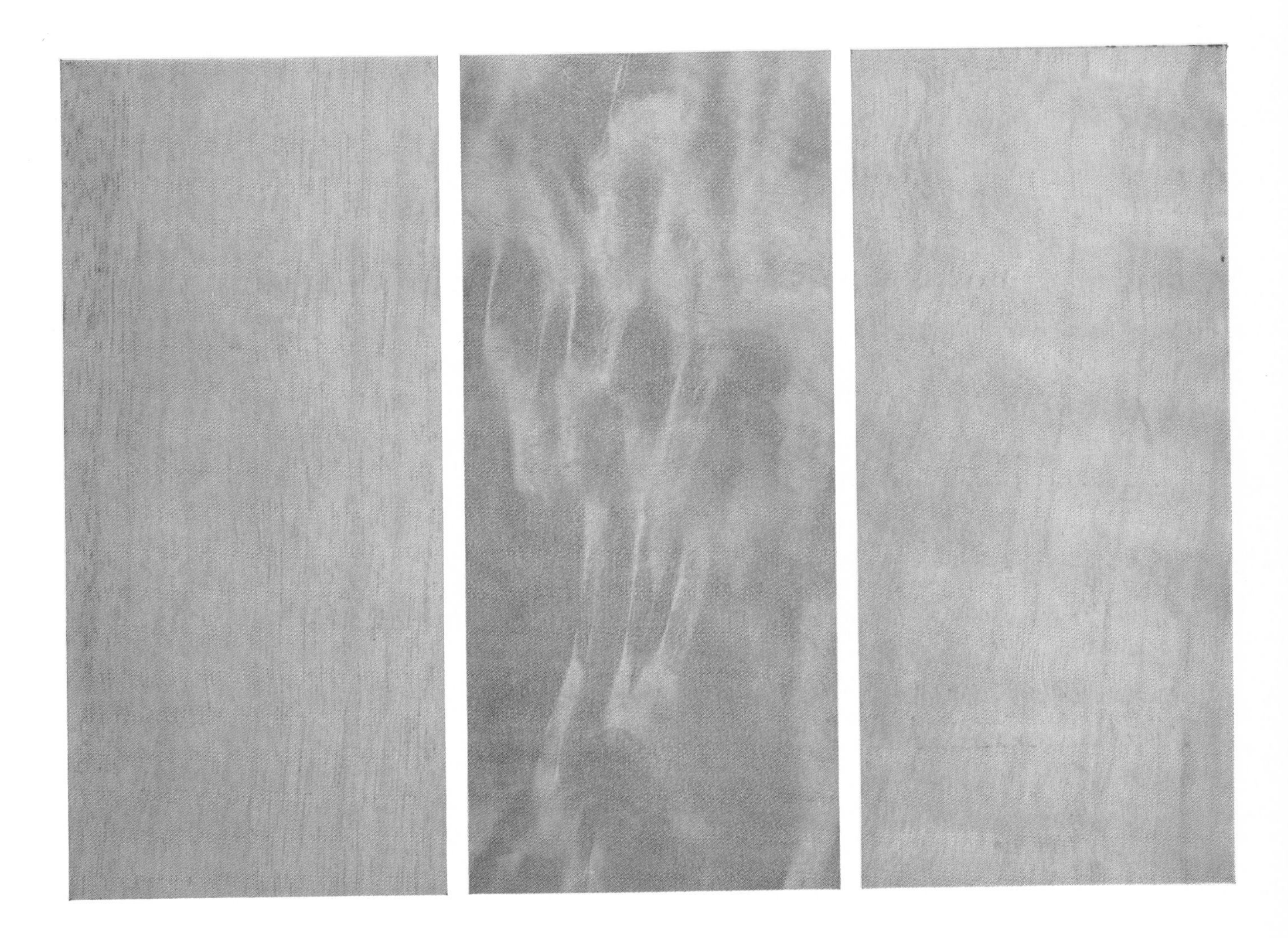

Avodiré

Quarter-cut

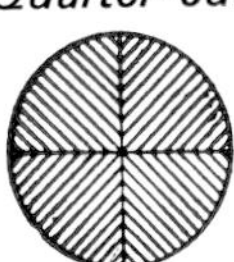

Curl

Butt

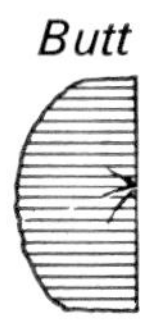

Reproduced actual size

Ayan

[*Distemonanthus benthamianus*]

The Nigerian name ayan has been adopted as the British Standard name. A more descriptive term is Nigerian satinwood—a tribute to its beauty of colour and grain. In France it is known as movingui.

Distribution and supplies. This species is widely distributed in the west African forest zone where it is reported to be abundant locally. It is a tall, rather slender tree; logs are upwards of 3·5 m. (say, 12 ft.) in length, mostly between 0·75 and 1·0 m. (30 and 40 in.) in diameter. It has been exported in relatively small quantities but can generally be obtained in dimensions suitable for flooring and joinery.

General description. A medium hardwood of striking appearance, bright-yellow (sometimes with darker streaks) with a fine texture and a bright, lustrous surface. The grain is often interlocked. Figured logs yield a highly decorative veneer.

Technical properties. Limited tests indicate that the timber can be seasoned with little degrade, and is fairly stable in service. It is about the same density as oak, and comparable to oak in strength. It is moderately durable under outdoor conditions and is reported to be resistant to subterranean termites. Working properties appear to be rather variable. The lighter-coloured, lower density material works fairly readily but the darker, denser timber causes rapid blunting of cutters, due to the silica in the wood, and can only be sawn satisfactorily with carbide-tipped saw teeth. Apart from its blunting effect, the timber works fairly readily and finishes cleanly if the cutting angle is reduced to 20°.

Uses. Ayan is of interest mainly for its decorative appearance and would seem to be specially suitable for use in the form of veneer. For decorative schemes it provides a good contrast with darker timbers. It makes a good floor for normal pedestrian traffic, and being more stable than most timbers, is suitable for use with subfloor heating. It is also recommended for gymnasium floors. Note that the wood contains a yellow substance, visible as a deposit in the pores, which can stain clothing, etc. under damp conditions.

Ayan

Flat-cut

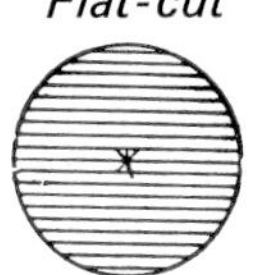

Reproduced actual size

Berlinia

[various species of *Berlinia*]

The botanical name *Berlinia* has been adopted as the trade name for the commercial species of this genus, including *B. confusa*, *B. grandiflora* and possibly others. The timber is sometimes sold under the name of rose zebrano, not to be confused with zebrano (see p. 176).

Distribution and supplies. The genus is widely distributed in West Africa. The timber has been exported in small quantities from Nigeria; logs of 0·5–0·9 m. (20–36 in.) diameter, 3·5–5·5 m. (12–18 ft.) long, are available for sawing to sizes required or for cutting into veneer.

General description. The heartwood has an unusually attractive appearance, pale to medium red-brown with somewhat irregular purple or brown streaks; it is well defined from the pale, featureless sapwood, often 100–150 mm. (4–6 in.) wide. The timber is about the same average density as European oak and, like oak, is variable in this respect. It is coarse textured, commonly with an interlocked and sometimes very irregular grain and the combination of colour variation and grain may give a highly decorative appearance, as in the illustration. The appearance is occasionally marred by dark-coloured streaks containing a hard gum, and brittleheart sometimes occurs in large logs.

Seasoning and movement. Berlinia seasons fairly slowly but well. It has a moderate shrinkage when dried from green to 12 per cent moisture content; once dry it has medium movement in service.

Strength and bending properties. It is approximately equal in weight to oak and slightly superior in all strength properties. It is classed as moderately good for steam bending.

Durability and preservative treatment. The heartwood is moderately resistant to decay, and reputed to be moderately resistant to termite attack; it resists preservative treatment. The sapwood is extremely susceptible to fungal and insect damage—particularly sap stain which may occur in drying—and to powder-post damage; it is permeable and readily treated.

Working and finishing properties. The working properties vary according to the density of the wood and general character of the grain. Reasonably straight-grained, light-weight material works fairly readily with machine and hand tools, with a moderate blunting effect on cutting edges; a tendency to tearing can be reduced by using a cutting angle of 20°. Denser wood is more resistant to cutting, with a marked dulling effect on cutting edges; wavy-grained wood may not machine satisfactorily even with cutting angles below 20°. The timber nails fairly well but tends to split near edges; except where very irregular grain is present the wood stains and polishes well and can be glued satisfactorily.

Uses. Until a few years ago berlinia was considered useful only for fairly heavy constructional work calling for strength and durability, as an alternative to oak and keruing, for example. More recently it has attracted attention as a decorative hardwood for high-class interior joinery in public buildings. Berlinia should be suitable for counter tops and fittings in banks and offices.

Berlinia

Quarter-cut

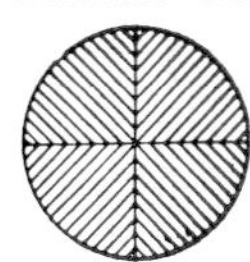

Reproduced actual size

Blackwood, African

[*Dalbergia melanoxylon*]

Distribution and supplies. Widely distributed in East and Central Africa. A small tree growing to a height of about 9 m. (30 ft.) and a diameter of about 20 cm. (8 in.). The timber has been exported for many years, from East African ports, in the form of small logs, 1–1·5 m. (say, 3—5 ft.) in length and variable in girth and quality.

General description. The heartwood is dark purplish-brown with black streaks which usually predominate so that the general effect is nearly black. It is exceptionally hard and heavy, density about 1·2 (75 lb./ft.^{3}), seasoned, with a fine, even texture and a slightly oily nature.

Seasoning and movement. The timber dries very slowly and tends to split in drying, especially in the log. The application of a moisture-retardant paint to the ends of the timber at every stage is advised to minimise splitting. Once seasoned, however, it is slow to absorb moisture.

Working and finishing properties. Considering its density, African blackwood has exceptionally good working qualities; it cuts smoothly and evenly, taking an excellent finish from the tool, as in turning and boring, and can be tapped for screw threads almost like metal.

Durability. The heartwood is believed to be very durable.

Uses. African blackwood is generally considered the best timber for ornamental turnery, being superior to ebony for this purpose. For wood-wind instruments, also, it is preferred to ebony because of its oily nature and resistance to climatic changes. With the exception of cocuswood no other timber has been found which equals African blackwood for reliability, working qualities and the tone of the finished instrument. The wood is also used for turned goods such as brush backs, knife handles, pulley blocks and chess-men.

Blackwood

Flat-cut

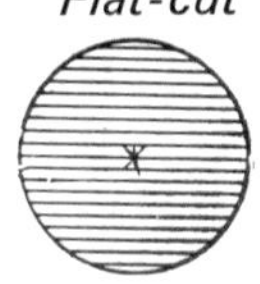

Reproduced actual size

Bubinga or Kévazingo

[three species of *Guibourtia*]

The timber of three closely allied species of *Guibourtia,* namely *G. demeusei, G. pellegriniana* and *G. tessmannii* (formerly classified as species of *Copaifera*) are similar in character and usually marketed together. These species are known as bubinga in Cameroon and as kévazingo in Gaboon, and both names are used in international trade. Kévazingo from Gaboon is considered to be more highly figured than bubinga from Cameroon, and the tendency is to use the name kévazingo for the highly figured timber, irrespective of origin. (A detailed account of Bubinga and Allied Timbers, by J. D. Brazier, was published in *Timber Technology and Machine Woodworking,* 1955, vol. 63, pp. 237–239.)

Distribution and supplies. The geographical distributions of the three species overlap. *G. tessmannii* is reported to be the principal source of bubinga from Cameroon while *G. pellegriniana* is more common in Gaboon. *G. demeusei* is more widely distributed, from South-East Nigeria, through Cameroon and Gaboon to the Congo region. The trees are fairly large, with clear boles 9–18 m. (30–60 ft.) long and 1–1·5 m. (say, 3–5 ft.) in diameter. The timber has been exported, from Cameroon and Gaboon, mainly in log form.

General description. The heartwood is a medium reddish-brown or purplish-brown with darker veining, similar to some types of rosewood but finer in texture. It is of the same order of density as Brazilian rosewood, between 0·80 and 0·96 (50 and 60 lb./ft.^{3}) in the seasoned condition. The grain may be straight or interlocked but some logs have very irregular grain which gives a highly decorative figure to rotary-cut veneer.

Working and finishing properties. Although relatively hard and dense the timber is reported to saw without difficulty, though the power consumption is fairly high. It takes a fine, clean finish.

Uses. It has been used mostly in the form of veneer for decorative purposes, and would seem to have further possibilities as an alternative to rosewood for high-class furniture and cabinet work, and fancy turnery such as knife handles and brush backs.

Other species of interest. *Allied species of* Guibourtia *furnishing decorative timbers are ovangkol* (*see p. 162*), *mutenye* (*see p. 146*) *and Rhodesian copalwood* (G. coleosperma).

Bubinga

Rotary-cut

Quarter-cut

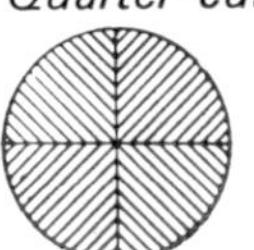

Curl

Reproduced actual size

Camwood or African Padauk

[*Pterocarpus soyauxii*]

Under the name of camwood or barwood this species was formerly well known as a dyewood, and is still used to a limited extent for this purpose. The trade name African padauk indicates its affinity with Andaman and Burma padauk. It is also known as corail.

Distribution and supplies. This species occurs in Nigeria, Cameroon, Gaboon and the Congo region. It is available in the form of roughly hewn billets up to about 1·25 m. (4 ft.) long and 0·3 m. (1 ft.) in diameter, and as logs, up to 1m (say, 3ft.) in diameter for conversion to veneer or sawn timber. The quantity exported is small but the timber can generally be obtained from hardwood specialists if required.

General description. The heartwood is remarkable for its deep-red colour and natural lustre. It has a moderately coarse texture due to the large pores. The grain is fairly straight. The density is between 0·65 and 0·80 (40 and 50 lb./ft.^{3}), seasoned.

Technical properties. Camwood seasons fairly slowly, with the minimum of degrade; its dimensional movement in service is exceptionally small. The timber is classed as very durable. Considering its density it is fairly easy to work with hand and machine tools, and like other species of padauk, takes an excellent finish.

Uses. It is used to a limited extent for fancy turnery and for the handles of high-quality tools such as chisels. It is an excellent flooring timber, suitable for heavy pedestrian traffic in public buildings, particularly where a good appearance is desired. By virtue of its stability it is acceptable for flooring where subfloor heating is installed. Though not often used for furniture or joinery it would be suitable for special work.

Camwood

Flat-cut

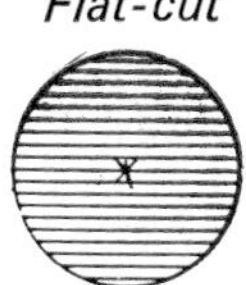

Reproduced actual size

Ceiba

[*Ceiba pentandra*]

On the international market this timber is known as fromager or fuma. Timber of the closely related species *Bombax buonopozense* is sometimes marketed with ceiba, mixed consignments being invoiced as ceiba/bombax.

Distribution and supplies. Ceiba is extremely common and widely distributed in the tropics, including West Africa and America. It is a large tree with a straight, cylindrical bole, commonly 12–15 m. (40–50 ft.) long and nearly 2 m. (6 ft.) in diameter above the large buttresses. In recent years it has been exported to Europe, mostly in log form, from the Ivory Coast, Nigeria and Cameroon and could probably be supplied by any of the timber-producing countries of West Africa.

General description. A very soft, coarse-textured, light-weight wood, similar to obeche but softer. It is variable in density, from 0·20 to 0·45 (13–28 lb./ft.^{3}) when dry, average about 0·32 (20 lb./ft.^{3}), which compares with obeche at 0·38 (24 lb./ft.^{3}). Ceiba shows no clear distinction between sapwood and heartwood; the natural colour varies from pale yellowish-brown to pinkish-brown but it is extremely prone to fungal discoloration. The grain is interlocked and sometimes irregular. The wood has a somewhat harsh appearance due, in part, to its coarse texture, and it lacks the high natural lustre and smooth feel of some other light-weight woods such as obeche and balsa.

Seasoning and movement. Ceiba is said to season rapidly and without much distortion; FPRL kiln schedule J is recommended. Once dry, ceiba is considered to be stable in use.

Strength and bending properties. Even allowing for its light weight, ceiba appears particularly weak in most strength properties. It is about 15 per cent lighter in weight than obeche but limited data suggest that it has only about two-thirds the strength of that timber in compression, bending and shock resistance. It is unsuitable for steam bending without special support.

Durability and preservative treatment. A perishable timber, very susceptible to both fungal and insect attack and requiring speedy extraction, conversion and drying to prevent rapid deterioration. Dry timber is sometimes attacked by powder-post beetles. It should not be exposed to conditions favouring decay without proper preservative treatment. It is very absorbent and readily impregnated with preservatives.

Working and finishing properties. Ceiba is a difficult wood to saw cleanly and finish smoothly because of its low density. Sawn surfaces tend to be woolly, and to finish to a smooth surface, cutters should be maintained in a sharp condition. Care is needed in boring, end-grain working and turning, and sharp tools are necessary to avoid tearing and breaking away. Sanding and scraping should be carried out with caution because of the softness of the wood. It takes nails and screws readily but has poor holding properties. Ceiba peels to give a good veneer provided that logs are fresh and free from insect and fungal attack. Veneers can be glued satisfactorily.

Uses. A light-weight timber, intermediate between balsa and obeche but rather coarser-textured and lacking their attractive, lustrous appearance. Successful utilisation depends on rapid extraction, conversion and drying. It has been used for packing cases and very light-weight joinery, but appears to be most suited for conversion to veneer, for plywood core stock. It is also suitable for insulation purposes.

Ceiba

Rotary-cut

Reproduced actual size

Cordia, West African

[*Cordia millenii* and *C platythyrsa*]

Distribution and supplies. *Cordia millenii* and *C. platythyrsa* are widely distributed in tropical Africa, *C. millenii* from the Ivory Coast to Uganda and Kenya, and *C. platythyrsa* from Sierra Leone to Gaboon. In Nigeria, *C. millenii,* which is believed to be the more important species, yields logs 3·5 m. (12 ft.) and up in length and 0·5–1 m. (20–40 in.) in diameter. The following information is based mainly on timber of *C. millenii; C. platythyrsa* is believed to be similar.

General description. West African cordia is a light-weight, fairly coarse-textured but lustrous wood, of pleasing appearance, pale golden-brown to medium-brown, occasionally with a pinkish tint, somewhat similar to teak. The grain tends to be irregular; it is typically interlocked to give a stripe which is further enhanced on accurately quarter-cut surfaces by a ray figure. The average density is about 0·43 (27 lb./ft.^{3}) when seasoned, which is somewhat heavier than obeche but lighter than agba and teak. Brittleheart is fairly common and is generally characterised by a pale colour and a lower than average density.

Seasoning and movement. It seasons rapidly and well, although a high-temperature schedule is necessary to accelerate moisture movement and remove localised patches of moisture. FPRL kiln schedule K is recommended. Once dry the timber has a small movement with changes of relative humidity.

Strength properties. Cordia is intermediate in strength between obeche and agba. It is about 12 per cent heavier than obeche and some 20 to 25 per cent stronger in bending and compression, about 40 per cent harder and 15 per cent stronger in shear; in its resistance to it is about the same as obeche.

Durability and preservative treatment. The timber's reputation for durability is partly borne out by the results of laboratory tests which have demonstrated that the denser wood is very durable when exposed to fungal attack while lighter-weight material gives a less satisfactory performance. The timber is reputed to be resistant to treatment by pressure methods.

Working and finishing properties. Cordia works and finishes easily and well, with only a slight blunting effect on tools. For rip-sawing, saws with 46 or 54 teeth and a 25° hook angle are recommended. In planing, a standard 30° block is satisfactory for plain-sawn surfaces but, in general, a reduction to 20° is recommended to avoid tearing of quarter-cut material. Boring, recessing and moulding can be carried out satisfactorily, although sometimes with a slightly fibrous finish and some tearing out on sides and ends. It can be nailed satisfactorily but fine gauge nails are recommended.

Uses. In West Africa cordia is rated a high-class furniture wood. Practical tests in Britain have confirmed that it has good working and finishing properties and stability in service. Its attractive appearance and superficial resemblance to teak and afrormosia suggest a suitability for joinery and cabinet work though because of its relatively low strength it is less acceptable for the structural parts of furniture or for hard wear. Its combination of stability and—with selection—durability suggests its use in boat building where its light weight could be an advantage. It is rather too light in weight and coarse-textured for flooring.

Other species of interest. Cordia goeldiana *of Brazil, known as freijo, has been used successfully in the USA for furniture and interior joinery and has been shipped occasionally to Britain in small quantities.* C. alliodora *(salmwood) of the West Indies and tropical America is somewhat similar.*

Cordia

Flat-cut

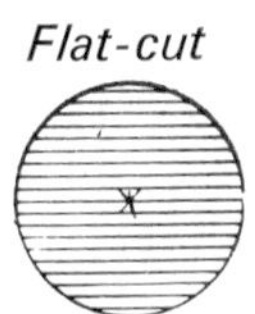

Reproduced actual size

Dahoma or Ekhimi

[*Piptadeniastrum africanum* formerly called *Piptadenia africanum*]

This timber is variously known as dahoma (Ghana), ekhimi or agboin (Nigeria) and dabema (Ivory Coast) according to the country of origin.

Distribution and supplies. It is widely distributed in West, Central and parts of East Africa. It grows to a height of 37 m. (120 ft.) or more, with a diameter of 0·9–1·25 m. (3–4 ft.) above the buttresses, but the utilisable length of clean bole is only 9–15 m. (30–50 ft.). Although it is in plentiful supply the timber is exported in relatively small quantities—mainly from Ghana—in the form of lumber.

General description. The yellowish-brown heartwood is somewhat similar to iroko in appearance, but less attractive. It is about the same density as iroko or oak. The grain is broadly interlocked, the texture rather coarse. The freshly cut timber has an unpleasant smell. In contact with iron and steel it is liable to become stained.

Seasoning and movement. It dries fairly slowly, with a marked tendency to collapse and distort. The movement of the seasoned timber in service is classed as medium.

Strength and bending properties. Dahoma compares favourably with oak and iroko in strength properties provided that it is used in fairly large dimensions. Owing to the strongly interlocked grain it is not suitable for use in small sections for purposes where strength is important. It is not recommended for steam bending.

Durability and preservative treatment. The heartwood resists the attack of subterranean termites, and is classed as moderately durable. It is rather difficult to impregnate with preservatives.

Working and finishing properties. On account of its fibrous nature, the wood blunts the teeth of mechanical saws fairly rapidly but generally has only a moderate blunting effect on other tools, and gives a relatively clean finish. It is inclined to tear in planing and moulding; a cutting angle of about 10° is recommended. It has good nailing properties and can be glued satisfactorily. The dust produced in machining operations causes irritation of the nose and throat.

Uses. Dahoma is best fitted for fairly heavy constructional work as an alternative to structural grades of oak.

Dahoma/Ekhimi

Flat-cut

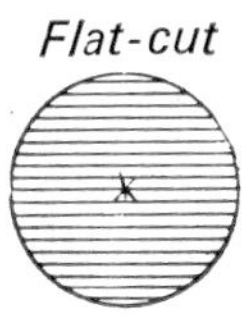

Reproduced actual size

Danta

[*Nesogordonia papaverifera* formerly called *Cistanthera papaverifera*]

Distribution and supplies. Danta is found in the principal timber-producing countries of West Africa. It is a tree of the mixed deciduous forests, of moderate size, the clean cylindrical bole 12–15 m. (40–50 ft.) long, 0·6–0·9 m. (2–3 ft.) in diameter. The timber has been exported in relatively small quantities from Nigeria and Ghana.

General description. A moderately heavy, high-class utility timber about the same density as oak and beech. Reddish-brown with an interlocked grain, suggesting sapele, and a fine, even texture. The appearance of the timber is often marred by pin knots.

Seasoning and movement. It dries rather slowly, with comparatively little degrade. Dimensional movement in service is classed as medium.

Strength and bending properties. Danta is particularly strong in bending, being superior to ash in this respect but inferior in shock resistance. It is classed as only moderately good for steam bending.

Durability and preservative treatment. The heartwood is reported to be resistant to fungal decay and termites, and is classed as moderately durable. It is resistant to preservative treatment.

Working and finishing properties. The timber works fairly readily with a blunting effect similar to that of beech. On quartered surfaces the grain tends to tear in planing; the cutting angle should be reduced to 15° to ensure a clean finish. In other operations the finish is generally satisfactory. The wood turns well, can be bored without difficulty but tends to split in nailing. Gluing is satisfactory.

Uses. Danta might be more widely used if its technical qualities were fully appreciated. It is eminently suitable for the framing and flooring of road vehicles, for telegraph cross-arms and garden furniture. It has a high resistance to abrasion and is recommended for flooring for heavy pedestrian traffic and for ballroom and gymnasium floors. In West Africa it is commonly used for boat building and for tool handles.

Danta

Quarter-cut

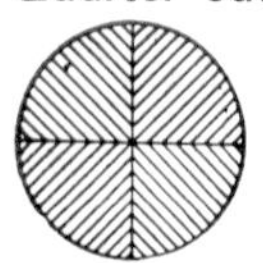

Reproduced actual size

Ebony, African

[various species of *Diospyros*]

The trade name ebony covers all species of *Diospyros* with predominantly black or streaked heartwood. Two of the principal species furnishing African ebony are believed to be *D. crassiflora* and *D. piscatoria*.

Distribution and supplies. Distribution of these species covers a wide area, including the Ivory Coast, Ghana, Nigeria, Cameroon and Gaboon. The larger trees attain a height of 21 m. (70 ft.) and a diameter of 0·75 m. ($2\frac{1}{2}$ ft.) or more. Supplies for the export market, mainly from Nigeria, Cameroon and Gaboon, are in the form of hewn billets, free of sapwood, 1–2·5 m. (3–8 ft.) long by 100–400 mm. (4–16 in.) in diameter.

General description. The wood of some species is greyish-white or pinkish, sometimes with a small central core of black heartwood. Others have a fairly large brown, black or streaked heartwood. Some of the best black ebony is said to come from Gaboon. In Nigeria three kinds of ebony are recognised: (*a*) Black ebony free from light-coloured spots or streaks; (*b*) King ebony, with black or dark reddish-brown irregular stripes; (*c*) Queen ebony, with black and light-coloured (mostly yellow or pale-orange) irregular streaks. The heartwood is extremely fine textured and very heavy, average density about 1·01 (63 lb./ft.^{3}) in the seasoned condition. The light-coloured sapwood is less heavy.

Technical properties. The outstanding technical characteristics of ebony are its hardness and fine, even texture. In small dimensions it seasons fairly quickly, with little tendency to split or distort. The dark heartwood is very hard to work and has a fairly considerable dulling effect on cutting tools. It is of a brittle nature but cuts less harshly than East Indian ebony. The timber finishes smoothly in most operations and takes an excellent polish but when the grain is irregular it tends to pick up in planing; the cutting angle should be reduced to 20°. The light-coloured sapwood is milder and somewhat similar in working properties to American persimmon; it is suitable for steam bending.

Uses. African ebony is used in the manufacture of musical instruments, in particular for the finger-boards of violins and cellos, also for inlaid work, handles for small tools and cutlery, for turnery, brush backs and carving. The light-coloured sapwood is reported to be used in Nigeria for the handles of tools and agricultural implements, gun butts, flooring blocks and body building for road vehicles.

Other species of interest. *Other commercial varieties of true ebony are obtained from Ceylon (East Indian ebony), Celebes (Macassar ebony), Borneo and elsewhere. The names coromandel, calamander and Andaman marblewood refer to ebony with grey or brown mottling.* D. virginiana *is the American persimmon, traditionally used for making golfclub heads and shuttles for the textile industry.*

Ebony

Flat-cut

Reproduced actual size

Ekki or Azobé

[*Lophira alata*]

Also known as bongossi, kaku and red ironwood.

Distribution and supplies. This is one of the principal timbers exported from Cameroon, and is of common occurrence in other parts of West Africa also. It is one of the larger rain-forest trees, with a bole up to 30 m. (100 ft.) long and a diameter of 1·5 m. (5 ft.) or more. The timber is exported in log form and as sawn timber in large dimensions suitable for piles and heavy construction.

General description. An extremely hard, heavy timber, density nearly 1·3 (80 lb./ft.^{3}) in the green condition, 0·96–1·12 (60–70 lb. ft.^{3}), seasoned. Dark reddish-brown with fairly conspicuous whitish deposits in the pores. Grain usually interlocked; texture coarse and uneven.

Seasoning and movement. A difficult timber to season; it dries slowly with severe splitting and distortion. Dimensional movement in service is classed as medium.

Strength properties. Extremely strong, in the same strength class as greenheart.

Durability and preservative treatment. Noted for its outstanding resistance to fungal decay, termites and marine borers, and classed as very durable. It is extremely difficult to impregnate with liquids but is not among timbers that call for preservative treatment.

Working and finishing properties. The timber is difficult to work, especially in the seasoned condition, when it has a severe blunting effect on cutting edges; logs should be converted—and resawn where necessary—before the timber has had time to dry. Though very hard, it can be planed to a fairly smooth finish. It cannot be nailed without pre-boring but glues reasonably well.

Uses. Being exceptionally strong and durable and obtainable in large dimensions, ekki is mainly used for heavy constructional work. On the Continent—as in the countries of origin—it has been widely used for piles, piers, jetties, harbour works, bridges, etc.; also for railway sleepers and wagon bottoms. It is a useful timber for wharf decking and for heavy-duty flooring, as in warehouses and factories.

Ekki/Azobé

Flat-cut

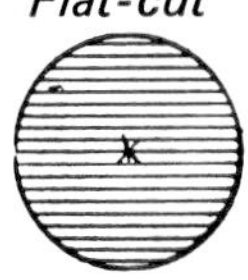

Reproduced actual size

Esia

[*Combretodendron macrocarpum*]

Esia or essia is the Ghanaian name for this species. Other names are abale (Ivory Coast), owewe (Nigeria), minzu (Congo).

Distribution and supplies. Widely distributed and of common occurrence in the West African rain forests. A tall tree with a cylindrical, unbuttressed bole, 0·6–0·9 m. (2–3 ft.) in diameter. Supplies are probably more than sufficient to meet the limited demand. It has been exported in log form for conversion to veneer.

General description. The pale to medium reddish-brown, relatively coarse-textured heartwood has a distinctive appearance, particularly on quartered surfaces where darker streaks and the conspicuous rays produce an attractive figure; the grain is sometimes straight, sometimes shallowly interlocked or, very occasionally, wavy. The sapwood is pale, commonly 50–75 mm. (2–3 in.) wide and susceptible to fungal stain and pinhole borer damage. When fresh the wood has a powerful, unpleasant smell but this does not persist once the wood is dry. It is moderately heavy, density about 0·70 (44 lb./ft.^{3}), seasoned, of the same order as oak.

Seasoning and movement. Esia dries slowly and is very prone to check, split and distort, and, even when air dried, thick stock may collapse and develop internal splits. It is doubtful whether the timber can be kiln seasoned satisfactorily from the green condition but for stock up to 38 mm. (1$\frac{1}{2}$ in.) thick FPRL kiln schedule B, modified to increase humidity by 10 per cent at every stage, may prove satisfactory. Shrinkage is high, especially in the radial direction, in which it is only a little less than the tangential shrinkage; once dry the timber is very susceptible to changes in humidity, and is classed as having a large movement.

Strength and bending properties. A strong timber, particularly in static bending and compression, but apt to break suddenly under impact. It is unsuitable for steam bending.

Durability and preservative treatment. The heartwood is resistant to fungi and termites but is sometimes damaged by pinhole borer attack; it is very resistant to impregnation but the sapwood, which is perishable, can be readily treated.

Working and finishing properties. Esia is relatively hard to work; it can be sawn and machined without undue blunting of tools but tends to cause vibration of saw teeth unless they have a fairly small pitch and only moderate amount of hook. Newly sharpened cutters give a good finish on planing but, in general, reduction of the cutting angle to 20° is necessary to prevent tearing of planed surfaces. There is some tendency to char on boring, and pre-boring is advisable to prevent splitting on nailing.

Uses. Although locally abundant in West Africa and with a good stem form, esia presents considerable difficulties in utilisation—mainly because of the degrade which occurs in drying. It may find a use in the countries of origin for heavy, rough construction work and possibly for railway sleepers, provided surface-and end-splitting are not considered objectionable. It is of uncertain value for rotary-peeled plywood but can be sliced to produce a decorative quartered veneer; figured logs may be of interest for this purpose.

Esia

Flat-cut

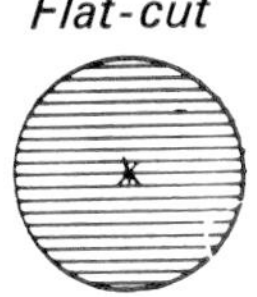

Reproduced actual size

Gaboon or Okoumé

[*Aucoumea klaineana*]

The trade name gaboon or gaboon mahogany is taken from the former French colony of Gaboon (Gabon) in equatorial Africa, the principal source of supply. The French name okoumé is also widely used in international trade. The use of the name Gaboon mahogany for this timber is confusing since true African mahogany (*Khaya ivorensis,* see p. 128) is also exported from Gaboon.

Distribution and supplies. This species is practically confined to Gaboon and the adjoining territories of Spanish Guinea and the former French Congo. Within this limited area it is extremely common. It is the principal timber exported from Gaboon (mainly to France, Western Germany, Israel and the Netherlands) and in terms of quantity it is the most important West African timber on the world market. The tree grows to a large size with a clear cylindrical bole, up to 21 m. (70 ft.) in length above the large wing-like buttresses. Logs for export vary from 0·6 to 1·8 m. (2–6 ft.) or more in diameter (average about 1·1 m. ($3\frac{1}{2}$ ft.), 3·5–11 m. (12–36 ft.) in length. The bulk of the timber is shipped in log form for the manufacture of plywood, with relatively small quantities of sawn timber. Plywood for export is also manufactured locally in Gaboon.

General description. The narrow sapwood is of a pale-greyish colour, the heartwood, salmon-pink darkening to pinkish-brown, with a general resemblance to African mahogany though somewhat lighter in colour and lighter in weight, average density about 0·43 (27 lb./ft.^{3}) in the dry condition. When cut on the quarter the grain is seen to be more or less interlocked but the surface of rotary-cut plywood presents a straight or, more often, a slightly wavy grain. Figured logs are reported to occur in the proportion of 3 or 4 per cent of the total production. The wood has no pronounced odour or taste.

Seasoning. Gaboon is not difficult to dry, either in a kiln or stacked in the open, and there is no marked tendency to distortion with conventional seasoning treatments.

Strength properties. The timber is fairly strong for its weight but is not often used for purposes where strength is important. It may be compared to poplar in strength properties.

Durability. It is not resistant to decay in situations favouring fungal development.

Working and finishing properties. The timber tends to be woolly, and blunts saw teeth fairly quickly. Sharp cutters are essential to obtain a smooth finish. It is almost ideal for conversion into veneer, either by peeling or slicing.

Uses. The bulk of the timber produced is manufactured into some form of composite wood: either plywood, laminboard or blockboard. These are used principally in the joinery and furniture industries, for the carcases of wardrobes, etc., flush doors, hidden parts of furniture, e.g. backs of chests, bottoms and sides of drawers, and similar purposes; also for small boats where weight is a consideration. Thin boards are cut for the manufacture of boxes for packing cheap cigars, in place of the true cigar-box cedar (*Cedrela* species). Figured logs are sliced to produce semi-decorative veneer.

Gaboon/Okoumé

Quarter-cut

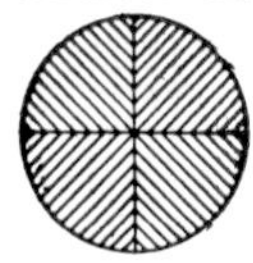

Rotary-cut

Reproduced actual size

Gedunohor, Tiama or Edinam

[*Entandrophragma angolense*]

This timber is variously known in the trade as gedunohor (Nigeria), tiama (Ivory Coast, Gaboon and Congo) and edinam (Ghana) according to the country of origin. As the botanical name indicates, it is closely allied to sapele (*Entandrophragma cylindricum*) and utile or sipo (*E. utile*) of the mahogany family (Meliaceae).

Distribution and supplies. Widely distributed in West and Central Africa; large supplies are available in the Ivory Coast, Ghana and Nigeria. A large, well-shaped tree, providing cylindrical logs 0·9–1·8 m. (3–6 ft.) in diameter, 3·5–9 m. (say, 12–30 ft.) in length. The timber is regularly shipped to Europe in large quantities, usually in log form but also as sawn timber, mainly from the Ivory Coast and Ghana. In terms of quantity exported it is less important than sapele, utile and African mahogany. Square-edged timber is stocked in a wide range of sizes: 25–100 mm. (1–4 in.) thick, 150–350 mm. (6–14 in.) wide, in lengths of 2–6 m. 6½–20 ft.). Logs are available for sawing to special lengths and sizes.

General description. A timber of the mahogany type. Compared with sapele and African mahogany it is of plain appearance, rather dull reddish-brown; it has the characteristic interlocked grain but well-figured wood is comparatively rare. Lighter in weight than sapele and utile, about equal to African mahogany in this respect, averaging 0·55 (34 lb./ft.^{3}) in the seasoned condition. It sometimes has a pleasant, cedar-like scent, though less pronounced than in sapele.

Seasoning and movement. There is a marked tendency to distort in seasoning. The dimensional movement in service is rated as small but is slightly inferior to that of African mahogany.

Strength and bending properties. Limited tests indicate little difference between gedunohor and African mahogany in strength. The timber is not so strong as sapele and utile. It appears to be unsuitable for steam bending.

Durability and preservative treatment. Like most timbers of the mahogany family, gedunohor is only moderately resistant to fungal and insect attack, and is difficult to impregnate with preservatives.

Working and finishing properties. It is fairly easy to work with machine and hand tools. The interlocked grain is liable to pick up on quarter-sawn surfaces unless the cutting angle is reduced to 15° or less. Otherwise it generally gives a satisfactory finish, provided that sharp cutters are used, and good results are achieved in staining, polishing and gluing.

Uses. This species is used mainly as a substitute for African mahogany, both in the solid and as veneer, for furniture and joinery; also for block and strip flooring. In furniture manufacture it is used for the cheaper lines of mahogany furniture, and for interior work such as drawer sides—on account of its small dimensional movement. The cylindrical shape of the logs makes it extremely suitable for rotary peeling for veneer production.

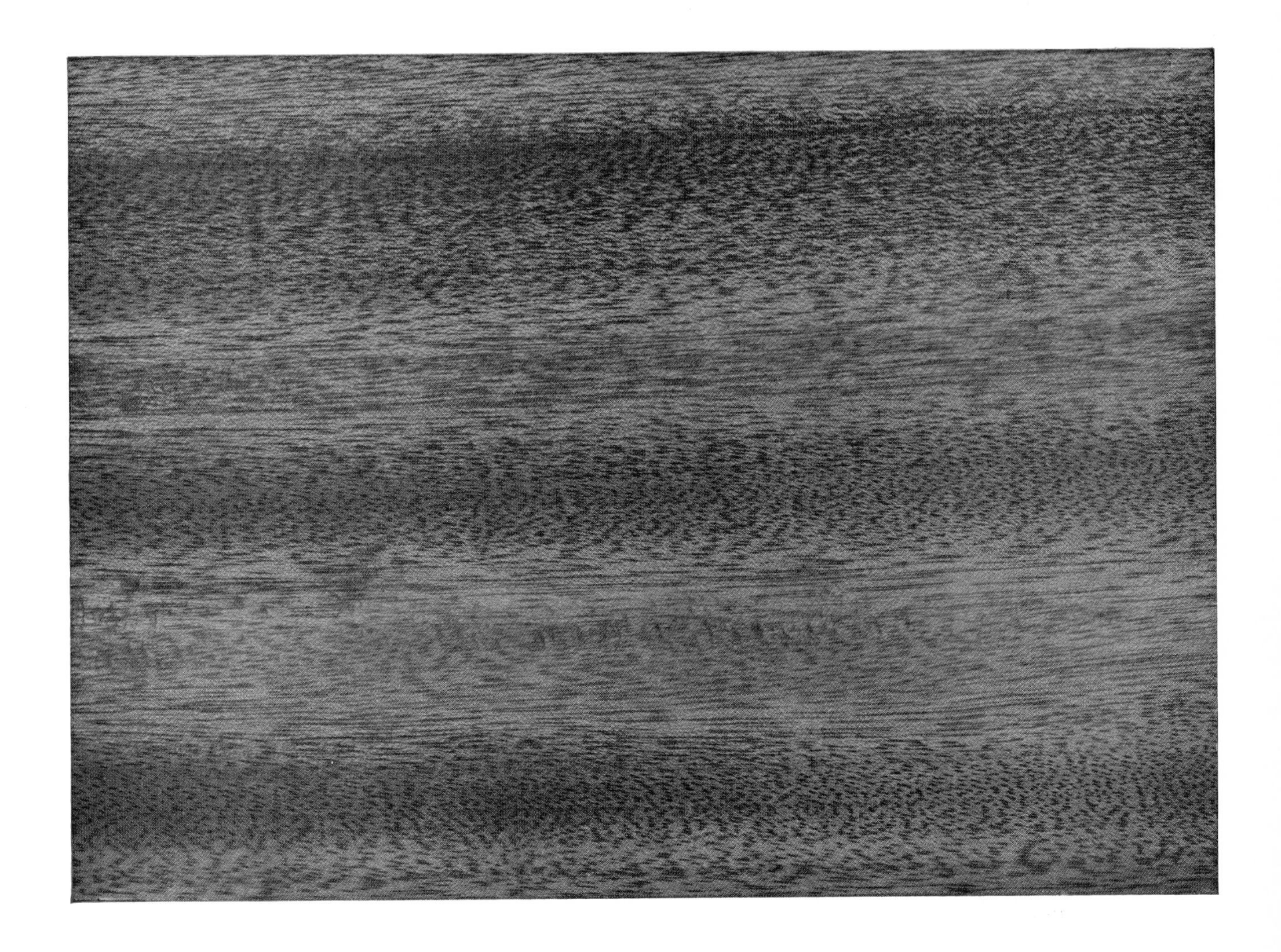

Gedunohor

Quarter-cut

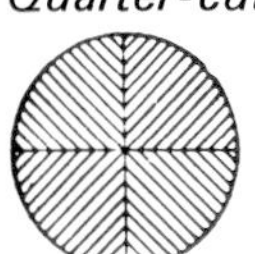

Reproduced actual size

Guarea

[*Guarea cedrata* and *G. thompsonii*]

These two West African species are generally marketed under the same trade name: guarea. Where they occur together, as in Nigeria, they may be shipped separately under distinctive names, white or scented guarea (*G. cedrata*) and black guarea *G. thompsonii*); the terms white and black refer to the bark of the tree, not to the wood. The French name for *G. cedrata* is bossé.

Distribution and supplies. The two species are widely distributed in the rain-forest zone of West Africa, *G. cedrata* being the more common. Both species are large trees, 0·6–1·25 m. (2–4 ft.) in diameter above the buttresses, though logs for export from Nigeria are usually of medium size, 0·5–1 m. (say, 20–40 in.) in diameter. The timber is exported in substantial quantities from the Ivory Coast and Nigeria, and in smaller quantities from other timber-producing countries in West Africa. It is available in log form for conversion as required, and is stocked as sawn timber, 25–75 mm. (1–3 in.) thick, 150–375 mm. (6–15 in.) wide, in lengths up to 4·5 m. (15 ft.), and as strips and squares; also as decorative veneer.

General description. These two mahogany-type timbers differ slightly in appearance and technical qualities. *G. cedrata* has a pinkish-brown colour with a pleasant, cedar-like scent which tends to disappear in time. It is somewhat firmer in texture, and denser than typical African mahogany, about 0·58 (36 lb./ft.^{3}), seasoned. Some logs are gummy. The grain may be either straight or wavy; a mottled or curly figure is not uncommon but the stripe figure characteristic of quartered sapele and African mahogany is not a regular feature. *G. thompsonii* is slightly heavier—density 0·63 (39 lb./ft.^{3})—of finer texture, and is said to be straighter in the grain.

Seasoning and movement. Both timbers can be dried fairly rapidly, without much degrade except for gum exudation, which can be objectionable. Movement in service is classed as small, though slightly more than that of African mahogany.

Strength and bending properties. Both timbers are somewhat harder and stronger than either African or Central American mahogany in most respects. Steam-bending properties are rated moderate to good.

Durability and preservative treatment. Resistant to dry rot but only moderately resistant to attack by other fungi. The heartwood is difficult to treat with preservatives.

Working and finishing properties. In working properties *G. cedrata* resembles African mahogany. The gum that comes to the surface in kiln drying is readily removed in machining and does not affect the finishing processes, though it has been known for gum exudation to occur later; a wax or oil finish has been recommended in preference to an impervious lacquer. *G. thompsonii* is considered to have superior working and finishing properties and has been likened to the denser grades of American mahogany. The dust produced in working guarea has been reported to cause skin irritation; there is some evidence that *G. thompsonii* is the offending species.

Uses. Guarea has been found satisfactory for furniture manufacture as an alternative to mahogany, notably for drawer sides, upholstery interiors and chair seats, and in reproduction work for its resemblance to Cuban and Central American mahogany. For flooring it is classed as suitable for normal conditions of pedestrian traffic, also for gymnasium and ballroom use, and for buildings where subfloor heating is installed. Guarea logs are reported to peel excellently, making a strong plywood of good appearance.

Guarea

Quarter-cut

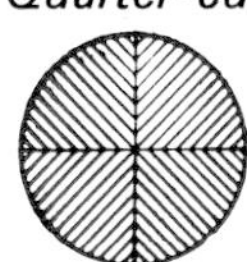

Reproduced two-thirds actual size

Idigbo

[*Terminalia ivorensis*]

The Nigerian name idigbo has been adopted as the British Standard name. The timber is also known as emeri (Ghana) and framiré (Ivory Coast), according to the country of origin. Note that the Nigerian name for the *tree*, black afara, referring to the colour of the bark, is liable to cause confusion with afara (*Terminalia superba*, see p. 78), and the use of this name for the *timber* is not recommended.

Distribution and supplies. Widely distributed in West Africa, though nowhere very abundant; the principal sources of supply are the Ivory Coast, Ghana and Nigeria. A tall tree with a bole free of branches for 20 m. (65 ft.) or more, straight but frequently fluted. Logs for export are 0·6–1·2 m. (24–45 in.) in diameter, 3·5–10 m. (12–33 ft.) long; the best quality timber is said to come from logs of medium size, up to 0·9 m. (35 in.) in diameter; larger logs are sometimes brittle in the heart. A timber of minor importance on the export market, but regularly available in log form or as sawn timber, 19–100 mm. ($\frac{3}{4}$–4 in.) thick, 75–375 mm. (3–15 in.) wide, 1–4·5 m. (3–15 ft.) long; also as decorative veneer.

General description. A pale-yellow to yellow-brown wood, moderately coarse-texured with a conspicuous growth-ring figure; flat-sawn idigbo has a superficial resemblance to plain oak. Quartered surfaces, unlike oak, have no ray figure but if interlocked or wavy grain is present, a decorative appearance results. Idigbo is somewhat variable in density; it averages about 0·55 (34 lb./ft.^{3}), seasoned, about the same as African mahogany, but lighter-weight wood is often associated with brittleheart which may also have a somewhat pinkish tint. The appearance of the timber is occasionally marred by wound tissue and gum veins following damage, probably by insects, to the growing tree.

Seasoning and movement. Idigbo seasons rapidly and well, with very little degrade. It has a small shrinkage on drying and is among the best hardwoods for its small movement in service.

Strength and bending properties. A moderately light-weight hardwood, idigbo has correspondingly low strength properties and is not rated a strong timber. If subject to stress, care must be taken to exclude material containing brittleheart. It is unsuitable for steam bending.

Durability and preservative treatment. Idigbo is unusual among light-weight pale woods for its high resistance to fungal attack; it is sometimes damaged by pinhole borer beetles, and the sapwood is susceptible to powder-post beetle attack. The heartwood is rated extremely resistant to preservative treatment.

Working and finishing properties. A mild wood which works fairly easily with all hand and machine tools. It finishes well, giving good results with stains and polishes after suitable filling. It has fairly good nailing and screwing properties, and is said to glue satisfactorily.

Uses. Idigbo is a good, all-round light hardwood combining an attractive appearance with good working and finishing properties, durability and outstanding stability. It is used for high-class joinery, for both interior and exterior work, and in furniture manufacture, mainly for framing. It is unsuitable for draining boards as it contains a yellow dye which is liable to stain clothes. When used out of doors, iron fittings—unless protected—should be avoided, to prevent staining of the timber and corrosion of the iron. As flooring, idigbo is too soft for industrial uses but makes an attractive and serviceable domestic floor.

Idigbo

Quarter-cut

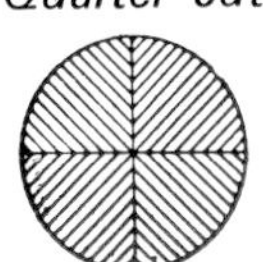

Reproduced actual size

Ilomba

[*Pycnanthus angolensis*]

The French trade name ilomba has been adopted as the British Standard name for this timber, which is also known as akomu (Nigeria), otie (Ghana) and walele (Ivory Coast).

Distribution and supplies. It is widely distributed, from Guinea, in West Africa, through the Ivory Coast, Ghana, Nigeria, Gaboon and the Congo region, to Angola in the south, and Uganda and Tanzania in the east. A medium-sized tree of excellent form, yielding long, cylindrical logs, 7·5–11 m. (25–35 ft.) in length and 0·6–0·9 m. (2–3 ft.) in diameter. The timber is exported in large quantities, almost entirely as logs, from West African ports, mainly to the Continent.

General description. A plain, pale-coloured wood of medium weight, similar in many respects to South American virola and East African mtambara (see p. 138). It is somewhat variable in colour, typically greyish-white to pinkish-brown, but occasionally with yellowish or mauve markings There is no colour contrast between the heartwood and the wide sapwood which is prone to discoloration if extraction and conversion are delayed. Ilomba is a straight-grained wood with a moderately coarse texture; density about 0·51 (32 lb./ft.^{3}) when dry, comparable to agba or a mild type of African mahogany.

Seasoning and movement. Ilomba requires careful and slow drying. It has a marked tendency to distort, split and collapse, and is especially difficult in thick sizes. It has an appreciable shrinkage, but once dry is reputed to be moderately stable when exposed to changes in relative humidity.

Strength properties. A wood with generally low strength properties; it is about average for its weight in bending strength and compression but tends to be somewhat brittle.

Durability and preservative treatment. A perishable timber requiring rapid extraction and protection to prevent serious degrade from insect and fungal attack. The sapwood is susceptible to powder-post beetle damage. In the untreated condition ilomba is unsuitable for use where exposed to fungal attack, but it can be readily treated with preservatives.

Working and finishing properties. A mild, straight-grained wood, ilomba presents no difficulty in sawing and planing, and takes a good finish. It nails and screws easily although with some tendency to split. It glues well, and is excellent for rotary peeling.

Uses. Ilomba is largely used for plywood production as an alternative to gaboon (okoumé). Sawn timber is used for interior joinery, for interior parts of furniture, mouldings, etc. On the Continent it is used for cigar boxes.

Ilomba

Flat-cut

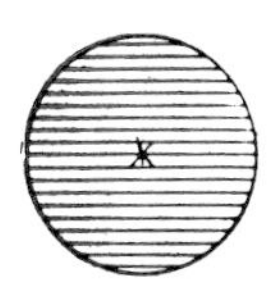

Reproduced actual size

Iroko

[*Chlorophora excelsa* and *C. regia*]

The Nigerian name iroko has been universally adopted as the trade name on the international market. In East Africa it is known as mvule.

Distribution and supplies. Iroko has a very wide area of distribution, extending across the continent of Africa from west to east. Production was developed during the second world war to meet the shortage of teak, and the wood is now exported by all the timber-producing countries of West Africa, except Ghana which has imposed a ban on exports of this timber; it is shipped from Kenya and Mozambique also. It is a large tree with a long, straight, practically unbuttressed stem which may be clear of branches to a height of 20 m. (65 ft.). Logs for export are 0·6–1·8 m. (2–6 ft.) in diameter, 4 m. (13 ft.) and up long, squared logs 0·45 m. (18 in.) square and larger. Square-edged timber is stocked in thicknesses of 16–100 mm. (⅝–4 in.), 75–600 mm. (3–24 in), or more, wide, 1–6 m. (3–18 ft.) long; it is also obtainable as veneer for decorative work.

General description. Iroko is somewhat variable in colour, from pale-yellow to medium-brown when fresh but soon darkening to a uniform brown colour. On exposure to the weather, as ships' decking, garden furniture, etc., it bleaches like teak. The timber is somewhat coarse-textured, with an interlocked and sometimes rather irregular grain, and of medium density, about 0·64 (40 lb./ft.^{3}), seasoned, which is about the same as teak. Iroko has a superficial resemblance to teak but is readily distinguished by its coarser texture, and lacks the characteristic greasy feel and leather-like smell of teak. An objectionable feature of iroko is the occasional hard calcareous deposits known as stone. These deposits occur in cavities associated with wounds in the wood and may not be discovered until the log is converted.

Seasoning and movement. The timber dries fairly rapidly, with only a slight tendency to splitting and distortion. Once seasoned its dimensional movement in service is very small, though some distortion is to be expected in long, narrow pieces with irregular grain when exposed to extreme variations of temperature and humidity.

Strength and bending properties. Iroko is harder than teak but slightly inferior in other strength properties. For steam bending it is classed as only moderate.

Durability and preservative treatment. In the green condition the timber is liable to damage from pinhole borer and longhorn beetles, but once dry the heartwood has a high measure of resistance to both insect and fungal attack. It cannot be effectively treated with preservatives.

Working and finishing properties. When free of stone, iroko presents little difficulty in working with hand or machine tools, although a 15° cutting angle is advocated when machining quartered surfaces if interlocked grain is present. Dulling of cutting edges is much less than with teak, unless stone is present, when it can become severe. Iroko can be nailed and screwed fairly well and takes stain but its coarse, open texture requires filling before polishing. It is not usually painted but can be if desired, using a thin primer.

Uses. In the countries of origin iroko is regarded as the local equivalent of teak, being outstanding in its lasting qualities and stability, and not unduly hard or heavy, with good all-round strength properties and a pleasing appearance. In Europe and North America its use has been developed largely as an alternative to teak; it has the advantage of being considerably less expensive. It is used, for example, in ship and boat building, for high-class joinery in public buildings, for laboratory benches, draining-boards, garden furniture and flooring. For exacting work it is advisable to eliminate pieces with irregular grain.

Iroko

Quarter-cut

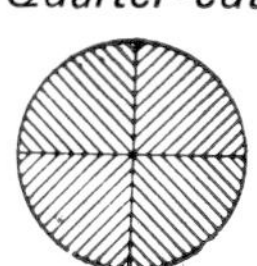

Reproduced two-thirds actual size

Mahogany, African

[various species of *Khaya*]

The general term African mahogany covers all species of *Khaya.* More specific names are sometimes used for the individual botanical species or for timber from a particular port or locality, as summarised below. Incidentally, it should be noted that since supplies of mahogany from the tropical American region have fallen off (species of *Swietenia* from Honduras, Nicaragua, Brazil, Peru, etc.), the term mahogany—without qualification—in specifications and descriptions of furniture, joinery, etc., generally means African mahogany.

West African mahogany is mainly the product of *K. ivorensis.* Shipments from certain areas may include *K. anthotheca* and *K. grandifoliola.* Trade names include Ivory Coast, Ghana, Nigeria, Grand Bassam, Degema, Takoradi, Lagos, Benin mahogany, etc., according to origin; also Lagos wood, Benin wood, etc. Acajou or acajou d'Afrique is the French trade name. In the USA it is sometimes sold as khaya to avoid confusion with Central American mahogany. *K. grandifoliola* is sometimes offered as *grandifoliola* or Benin mahogany.

Note that the term Sapele mahogany or Sapele wood, originally applied to mahogany-type timber shipped from the Nigerian port of Sapele is now generally restricted to the timber of *Entandrophragma cylindricum* (see p. 168). The term Gaboon mahogany for the well-known plywood timber *Aucoumea klaineana,* from Gabon or Gaboon in Equatorial Africa, has been dropped in favour of gaboon or the French trade name okoumé.

Distribution and supplies. *K. ivorensis* is a species of the coastal rain forests of West Africa. When fully grown the tree typically has a clean cylindrical bole up to about 25 m. (say, 80 ft.) in length above the buttresses. *K. anthotheca* and *K. grandifoliola* grow in regions of lower rainfall, further away from the coast; these species extend to Uganda and Tanzania. African mahogany is exported in log form or as sawn timber from all the timber-producing countries of West Africa, the principal sources of supply being the Ivory Coast, Ghana and Nigeria. Small quantities of sawn timber have been shipped from East Africa. Logs for export are 0·6–1·5 m. (2–5 ft.) in diameter and 3·5–9 m. (12–30 ft.) in length. Sawn timber is available in a very wide range of sizes, 16–100 mm. ($\frac{5}{8}$–4 in.) thick, up to 750 mm. (30 in.) wide, in lengths up to 6 m. (20 ft.), occasionally longer. Plywood is manufactured for export in Ghana, Nigeria and the Ivory Coast, in various European countries and in Israel. There is a big trade in veneers also.

General description. The timber varies in quality with the locality of growth. At best, it is only slightly inferior to Central American mahogany of average quality but on some soils the trees develop spongy hearts (brittleheart), and timber from these is apt to distort badly in drying and is difficult to finish smoothly with a firm surface. The colour varies from a light pinkish-brown to a deep-reddish shade; it seldom shows the yellowish-brown colour characteristic of the lighter shades of Central American mahogany. African mahogany generally is of somewhat coarser texture than the Central American wood. The density in the seasoned condition is about 0·53 (33 lb./ft.^{3}). The grain may be reasonably straight, but is more often interlocked, yielding striped or roe-figured boards when quarter-cut. Selected logs yield figured wood in great variety—crotch, swirl, mottle, fiddle-back, etc. A common defect in the timber is the presence of cross-fractures, called thunder-shakes or cross-breaks in the trade; this is frequently associated with the abnormal timber from trees with spongy hearts. There appears to be no significant difference between *K. ivorensis* and *K. anthotheca* so far as the appearance of the timber is concerned, and they are not usually separated in practice. *K. grandifoliola,*

[*continued on page 130*]

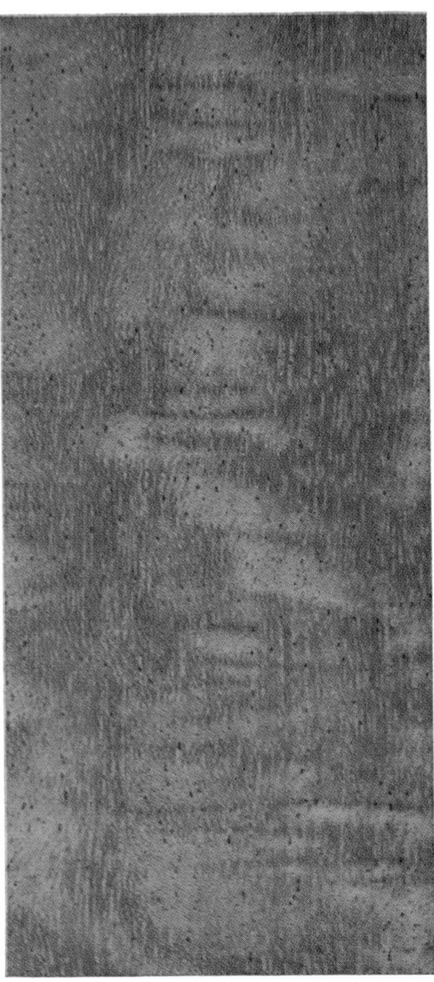

Mahogany

Flat-cut

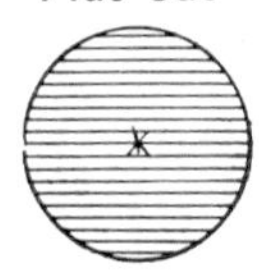

Quarter-cut

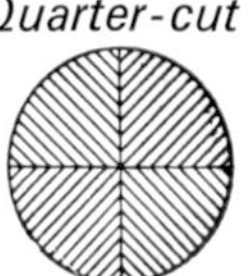

Rotary-cut

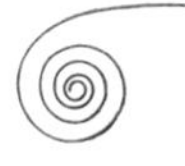

Reproduced actual size

MAHOGANY [*continued from page 128*]

however, is typically darker in colour, and appreciably denser, averaging about 0·70 (44 lb./ft.^{3}) and finer textured, resembling Cuban mahogany in general character.

Seasoning and movement. Seasoning presents no special problems, apart from the tendency of brittleheart to shrink unduly and unevenly. The dimensional movement of seasoned timber in service is rated as small. Test figures show that it is not quite so stable as Central American mahogany, but if clear, straight-grained timber is selected there is little to choose between the two species in this respect.

Strength and bending properties. The strength properties agree fairly closely with those of Central American mahogany but the African timber is more resistant to splitting and indentation. It is unsuitable for steam bending.

Durability and preservative treatment. African mahogany is fairly resistant to attack by wood-rotting fungi but in the log is susceptible to damage by longhorn beetles and pinhole borers. Sapwood is often attacked by powder-post beetles. The heartwood is very resistant to impregnation with preservatives.

Working and finishing properties. The timber from different districts varies in hardness, and this is reflected in the working qualities. On the whole it works without difficulty with both machine and hand tools, but when the grain is markedly interlocked a modified technique is desirable in planing and moulding operations. Brittleheart is apt to give a woolly surface from the tools. This is to some extent overcome by the use of sharp, thin cutting edges. The wood holds glue well and responds to modern finishing treatments satisfactorily. It is not usually painted but, if required, normal methods of painting can be used.
In common with certain other timbers of the mahogany family, African mahogany is liable to react with iron and iron compounds under damp conditions, resulting in dark stains on the wood. To prevent this, galvanised or non-ferrous metal should be used for fastenings and fittings.

Uses. African mahogany has largely replaced Central American mahogany as a standard timber for furniture, high-class joinery and fittings, boat building and many other purposes. Apart from its decorative appearance, it is valued for its technical qualities, notably its excellent working properties and stability in service. It has the further advantage of being readily obtainable in a wide range of sizes at a competitive price.
The timber of *K. grandifoliola,* being considerably harder and heavier than the general run of African mahogany, is more suitable where a stronger, harder-wearing wood is required: as for counter tops and block flooring.

Other species of interest. *Small quantities of the dry-zone mahogany,* K. senegalensis, *have been exported from Portuguese Guinea and marketed under the names of Guinea mahogany, bissilon and bissilongo. This species is widely distributed in the savanna forests of Africa, from Senegal to the Congo in the west, and across the continent to the Sudan and Uganda, outside the limits of the areas normally worked for export timber. The tree is comparatively small, such that it would be difficult to obtain dressed logs more than 450 mm. (18 in.) square. The wood is denser and harder than that of any other species of African mahogany, of the order of 0·80 (50 lb./ft.^{3}), seasoned, and the grain is inclined to be irregular so that it is more difficult to work. It is also distinguished by the dark reddish-brown or purplish-brown colour; in grain, texture and general appearance it closely resembles Cuban mahogany. Shipments to Europe have included some timber of excellent quality suitable for special purposes where a high-grade mahogany of firm texture and a natural dark colour is required.*
Another species of minor importance is K. nyasica *of East and Central Africa. The timber is exported from Mozambique—mainly to South Africa and Rhodesia—under the name of mbaua, umbaua or Mozambique mahogany. It tends to be darker and denser than typical West African mahogany.*

Dry–Zone Mahogany

Quarter-cut

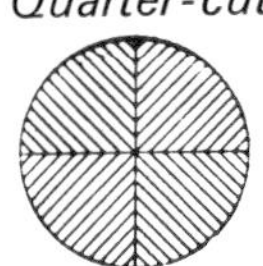

Reproduced actual size

Makoré

[*Tieghemella heckelii* formerly called *Mimusops heckelii*]

The Ivory Coast name makoré has been adopted as the international trade name. The timber has also been marketed as cherry mahogany in Britain, and as African cherry in the USA. Supplies from Ghana are sometimes invoiced as baku.

Distribution and supplies. This species is widely distributed in the mahogany districts of West Africa from Sierra Leone to Gaboon. It is an important export timber, shipped in large quantities from the Ivory Coast and Ghana, and in smaller quantities from other countries. It is a large tree with an exceptionally long, clean cylindrical bole, providing well-shaped, sound logs, 0·6–1·25 m. (2–4 ft.) in diameter, 3·5–7·5 m. (12–24 ft.) long. Square-edged timber is mainly 25–75 mm. (1–3 in.) thick, 150–375 mm. (6–15 in.) wide, available in lengths up to 4·5 m. (15 ft.). Figured logs are cut into veneer.

General description. Though not closely related to the true mahoganies, makoré has a superficial resemblance to a close-grained mahogany, being reddish-brown with a natural lustre, and compares favourably with mahogany in technical properties. It is appreciably denser than the general run of African and American mahogany, about 0·64 (40 lb./ft.^{3}), seasoned. Unlike mahogany, the wood is normally fairly straight in the grain; selected timber has an unusual mottled or chequered appearance when cut on the quarter, sometimes marked with irregular veins of darker colour. Makoré is liable to stain in contact with iron under damp conditions.

Seasoning and movement. The timber dries at a moderate rate, with only slight degrade due to distortion and splitting. Dimensional movement in service is small.

Strength and bending properties. Makoré is stronger than the general run of African mahogany, being more like sapele in this respect. Straight-grained timber is classed as moderately good for steam bending.

Durability and preservative treatment. It is resistant to attack by fungi and insects, including termites, and is classed as very durable. The heartwood is also extremely resistant to impregnation with liquids but it is not a timber which requires preservative treatment.

Working and finishing properties. Makoré causes rapid blunting of ordinary steel cutters and saws, presumably because of the silica in the wood. Saw teeth tipped with tungsten carbide are recommended for cutting seasoned timber. In other respects it works fairly well and takes a good finish. It can be glued satisfactorily but tends to split when nailed. The fine dust produced in some machining operations, and in sanding, is liable to irritate the nose and throat. It can be painted if required; the primer should be well thinned.

Uses. Makoré is a versatile timber. It is used for high-class furniture, joinery (including exterior joinery) and decorative work, both in the solid and as veneer, and is also valued for purposes where strength, durability and stability are important, including ship and boat building and the framing of motor vehicles; for structural plywood for exterior purposes and boat building. Other uses include rollers for the textile industry, miscellaneous turnery and flooring.

Other species of interest. Douka (*Mimusops africana)* and moabi (*Baillonella toxisperma* or *Mimusops djave)* are similar to makoré and frequently figured.

Makore

Quarter-cut

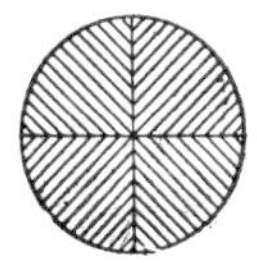

Reproduced two-thirds actual size

Mansonia

[*Mansonia altissima*]

Distribution and supplies. Mansonia was introduced to the world market from Nigeria, as a substitute for walnut, in the 1930s. It is regularly exported, mainly from Nigeria, Ghana and the Ivory Coast, mostly in the form of round logs. By West African standards the tree is of only moderate size. Logs are generally 500–760 mm. (20–30 in.) in diameter, 3·5 m. (12 ft.), and up, long. Square-edged timber is available in the usual thicknesses, 25-75 mm. (1–3 in.), in fairly long lengths but relatively small widths.

General description. The heartwood resembles American black walnut; it varies in colour from mid-brown to nearly black, often with peculiar tinges of red, purple or greyish-green. When freshly cut the colour is often very rich but tends to fade on exposure, to a somewhat indefinite brown, especially under the influence of sunlight. The sapwood is nearly white in colour. The wood is of a uniform, medium texture and the grain is usually fairly straight but some logs when opened show a striped or, more rarely, an irregular figure near the heart. The average density is approximately 0·6 (37 $lb./ft.^3$), seasoned, about the same as walnut.

Seasoning and movement. The timber can be dried fairly rapidly, with little distortion. Its dimensional movement in service is medium.

Strength and bending properties. On the basis of very limited tests it is rated as harder and more resistant to bending stresses than American black walnut but similar in other respects. It is a fairly good timber for steam bending, though somewhat variable in its behaviour, and should be suitable for making bends of moderate radius of curvature.

Durability. Mansonia is extremely resistant to fungal decay and is reported to be fairly resistant to termites. In the log it is susceptible to pinhole borer and longhorn beetle damage, but this is chiefly confined to the sapwood.

Working and finishing properties. Mansonia works easily with all hand and machine tools, and does not rapidly dull their cutting edges. It is easier to cut than American black walnut and, unlike the latter, has no marked tendency to char when worked on end grain. A good finish is obtained with standard conditions of working. It has fairly good nailing and screwing properties, takes stain, polish, etc. well, and can generally be glued satisfactorily. The dust produced in working and sanding mansonia is irritating to the eyes, nose and throat; efficent extraction plant and the use of protective masks is advisable to minimise this effect.

Uses. In Europe mansonia is mainly used in the solid, but also as veneer, as a substitute for walnut in furniture, cabinet work and joinery. It would be more widely used in this way but for the tendency to fade and for the irritating effect of the dust produced in manufacture.

Mansonia

Quarter-cut

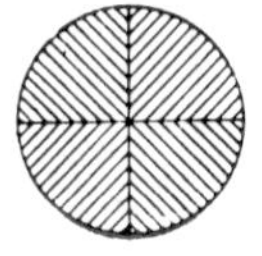

Flat-cut

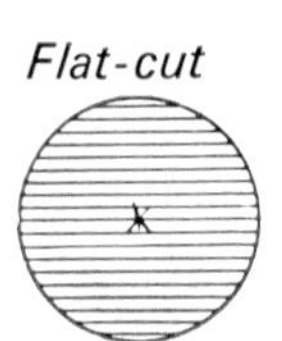

Reproduced actual size

Missanda or Tali

[*Erythrophleum guineense* and *E. ivorense*]

This timber is variously known as missanda (Mozambique), tali (Ivory Coast), erun or sasswood (Nigeria), potrodom (Ghana), kassa (Congo) and muave (Zambia), according to the country of origin.

Distribution and supplies. It is widely distributed in tropical Africa, as indicated above. A large tree up to 37 m. (120 ft.) high and 1·5 m. (5 ft.) in diameter, with an irregular bole. The timber is exported mainly in the form of flooring strips.

General description. The dark heartwood is extremely hard and heavy, density 0·90(56 lb./ft.^{3}), seasoned, with a fairly coarse texture and heavily interlocked grain.

Technical properties. Missanda is somewhat refractory in seasoning and not particularly stable. It is reputed to be extremely strong, and highly resistant to decay and insect attack (including termites) and marine borers. Being exceptionally dense and cross grained it is hard to work but finishes well, taking a fine polish.

Uses. It is used locally for heavy construction, such as bridges and harbour work, and for railway sleepers. In Europe it has been used successfully for heavy-duty flooring, as in schools, warehouses and factories.

Missandа/Tali

Flat-cut

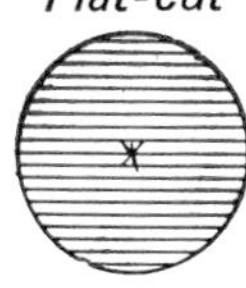

Reproduced actual size

Mtambara

[*Cephalosphaera usambarensis*]

Distribution and supplies. A large tree with a straight, cylindrical bole, 15–25 m. (50–80 ft.) in length and 1·25–1·8 m. (4–6 ft.) in diameter, above well-developed buttresses. It occurs in Tanzania and has been exported as lumber.

General description. Botanically this species is closely related to virola, and although somewhat heavier, the timbers are so similar in general appearance that it is not always possible to distinguish them. Mtambara is pale pinkish-brown with a faint orange tint; there is little colour contrast between heartwood and sapwood, although the latter tends to be somewhat paler. It has a generally straight grain and, with a moderately fine, even texture, is a rather featureless wood. Its density is about 0·60 (37 lb./ft.^{3}) when dry, that is a little heavier than the general run of African mahogany. It has a general resemblance to South American virola and West African ilomba (see p. 124).

Seasoning and movement. The timber air seasons slowly but well. It kiln dries quickly but with a risk of severe case hardening, which can be relieved without difficulty by a steaming treatment, and with some tendency to cup. When dry, movement of the timber is rated as medium when exposed to changes in relative humidity.

Strength properties. The strength of mtambara is about average for its weight in bending, compression and cleavage, above average in stiffness and shear strength, but rather low in hardness and resistance to impact.

Durability and preservative treatment. Although tests have not been made, it is likely that the timber has a low durability rating under temperate conditions; in East Africa it is perishable. Sapwood is extremely prone to fungal discoloration, and to obtain clean timber, logs are extracted and converted quickly and the timber given an anti-stain treatment. The wood is moderately resistant to impregnation but can be given an adequate preservative treatment by pressure methods.

Working and finishing properties. The timber saws easily and presents no difficulty in working with either machine or hand tools; it can be sanded to a smooth finish and takes polish well. It nails easily and holds nails well.

Uses. Mtambara is a general-purpose, utility timber. It has a plain appearance and, with a combination of moderately light weight, pale colour, straight grain and good working properties, is a useful wood for interior joinery and in the furniture industry. It peels well and is used for the manufacture of plywood in East Africa.

Mtambara

Quarter-cut

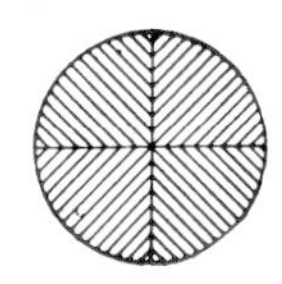

Reproduced actual size

Muhimbi

[*Cynometra alexandri*]

Distribution and supplies. Muhimbi is a common timber tree in Uganda and occurs in Tanzania and the Congo also. It has a clear bole up to 12 m. (40 ft.) long and 0·75 m. ($2\frac{1}{2}$ ft.) in diameter. The timber has been exported in the form of strips for the manufacture of flooring. Considerable quantities are available for purposes which warrant the fairly heavy costs of transport and manufacture.

General description. A very hard, dense, fine-textured timber, averaging about 0·90 (56 lb./ft.^{3}) in the seasoned condition. Generally of plain appearance, dull reddish-brown, with darker markings. The grain is inclined to be irregular.

Seasoning and movement. It dries slowly, with a strong tendency towards surface checking and end splitting but distortion is generally slight. Dimensional movement in service is classed as medium.

Strength and bending properties. The strength properties are outstanding; in most respects it is as strong, or stronger, than Canadian hard maple. It is classed as moderately good for steam bending.

Durability. Muhimbi is reputed to resist fungal and insect attack, including termites.

Working and finishing properties. Being exceptionally hard and dense it is difficult to work and has a fairly severe blunting effect on the cutting edges of tools. In planing, a satisfactory finish is generally obtained if the cutting angle is reduced to 15°. It is advisable to pre-bore for nails.

Uses. In the countries of origin muhimbi is used for heavy construction where strength and durability are required, including fender piles 300–350 mm. (12–14 in.) square for harbour work. In Britain it has been found suitable for strip and block flooring for heavy pedestrian traffic and for industrial purposes, as in factories, workshops and warehouses. It is also recommended for ballroom floors.

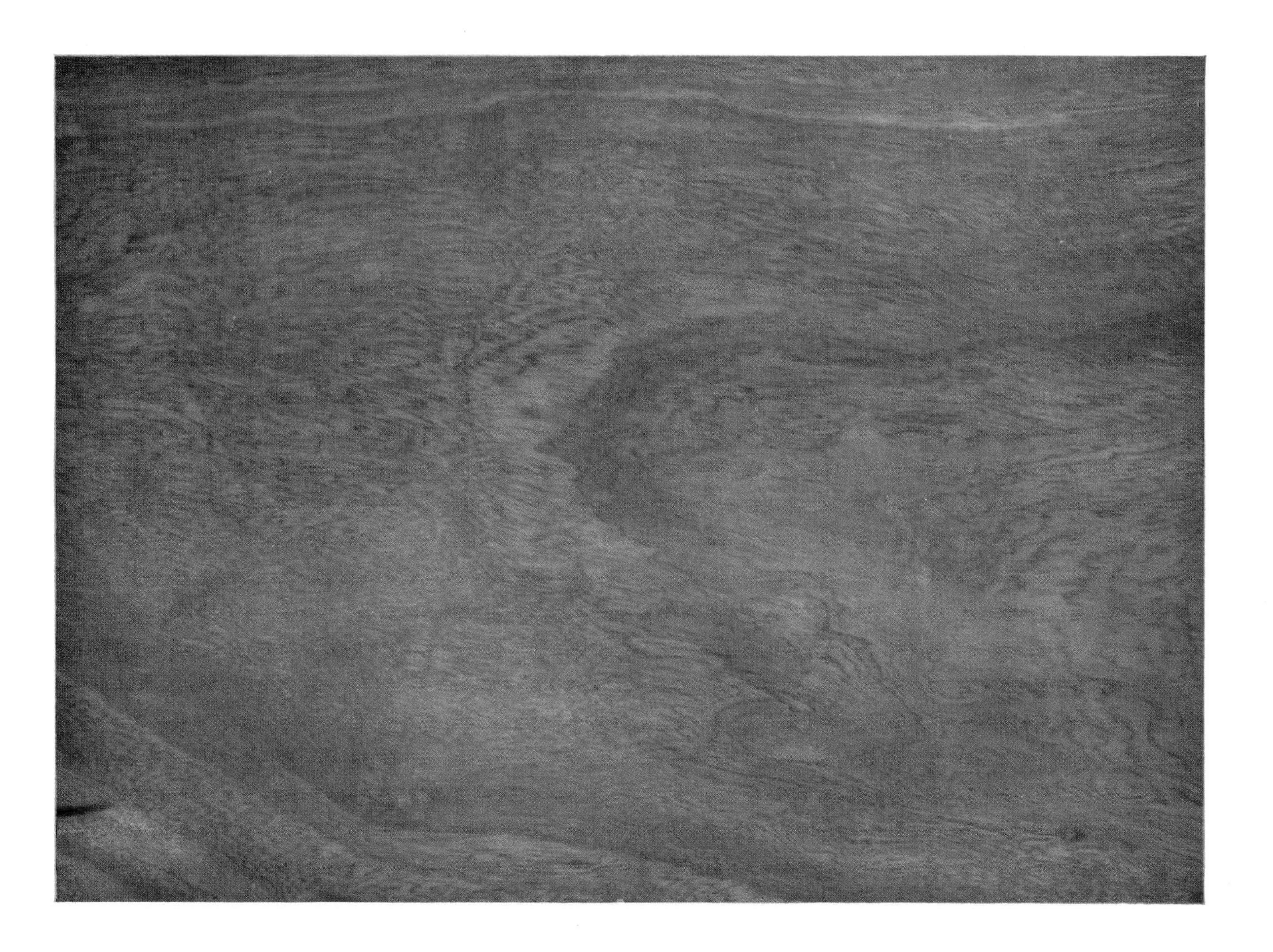

Muhimbi

Flat-cut

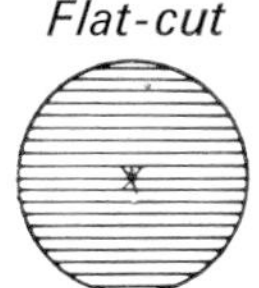

Reproduced actual size

Muhuhu

[*Brachylaena hutchinsii*]

Distribution and supplies. An East African timber, common in parts of Tanzania and Kenya. A small, often misshapen tree, yielding logs 2–6 m. (6–20 ft.) long with average diameter 450 mm. (18 in.). Muhuhu is one of a number of East African timbers exported to Britain for the first time after the second world war, when it was immediately recognised as a flooring timber of the highest class. It is shipped mainly as flooring blocks and strips.

General description A very hard, heavy timber with fine, even texture and interlocked grain, dark yellowish-brown, with a superficial resemblance to satinwood. Density about 0·91 (57 lb./ft.^{3}) in the seasoned condition, about the same as Rhodesian teak. When fresh the wood has a pleasant, spicy smell which is also noticeable when the dry wood is machined.

Seasoning and movement. Using FPRL kiln schedule B, 1 in. stock can be dried fairly rapidly, with only slight degrade due to surface checking and end splitting; in thicker sizes drying is slow and surface checking sometimes excessive. Once dry the timber is noteworthy for a small movement in service.

Strength properties. Although a heavy timber, muhuhu does not have outstanding strength, being generally inferior to beech in bending, stiffness and resistance to suddenly applied loads. Its resistance to indentation, combined with a fine texture, makes it highly resistant to abrasion.

Durability and preservative treatment. Heartwood is very durable under conditions favouring fungal attack, and is reputed to be resistant to termites and marine borers. It is rated very resistant to preservative impregnation using pressure treatment.

Working and finishing properties. Being very dense, with an interlocked and sometimes irregular grain, the timber presents some difficulty in sawing and finishing. For rip-sawing, a saw having 40 spring-set teeth with 15° hook is recommended to avoid overheating. Flat-sawn material planes satisfactorily under standard conditions, but for quarter-sawn stock a reduction of cutting angle to 20° is recommended. The timber requires to be pre-bored for nailing.

Uses. Attractive in appearance, hard wearing and extremely durable, muhuhu is outstanding as a flooring timber and is suitable for the most exacting conditions, as an alternative to Canadian hard maple. Its pleasing appearance makes it suitable for high-quality floors in hotels, public buildings and similar situations where there is heavy pedestrian traffic, and its excellent technical characteristics recommend it for flooring in factories, warehouses, etc.
Although technically suitable for a wide range of heavy construction work, its use for these purposes is limited by the relatively small sizes available. It is used in East Africa for carving and turnery, and an aromatic oil is extracted from sawmill waste for use in perfumery and toilet preparations.

Muhuhu

Quarter-cut

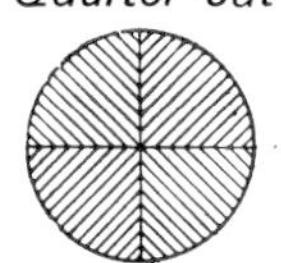

Reproduced actual size

Muninga

[*Pterocarpus angolensis*]

Muninga or mninga, the local name in Tanzania, has been adopted as the trade name in Britain. Elsewhere it is known as ambila (Mozambique), mukwa (Zambia) and kiaat or kajat (South Africa).

Distribution and supplies. A small to medium-sized tree, widely distributed in dry-forest areas from East Africa to the northern parts of South Africa. At maturity it is seldom more than 15 m. (50 ft.) high and 0·6 m. (2 ft.) in diameter, with a usually straight, clear bole of 7·5 m. (25 ft.). The qualities of this valuable timber are appreciated in the countries of origin so only limited quantities are exported to Europe. It is available in Britain in the form of logs, average about 350 mm. (14 in.) diameter and 2–2·5 m. (7–8 ft.) long, for veneer production and as square-edged lumber, 25–75 mm. (1–3 in.) thick, 150–300 mm. (6–12 in.) wide, in lengths of 2–5 m. (6½–16 ft.), and as flooring strips.

General description. A high-quality timber of handsome appearance, closely allied to Andaman and Burma padauk, amboyna and camwood or African padauk (see p. 100). The heartwood is basically a light to medium yellowish-brown with darker streaks, toning down to a pleasing, golden-brown shade. Its decorative character is sometimes enhanced by irregularly interlocked grain. The appearance of the wood is sometimes marred by localised chalky white deposits in the grain. The density in the seasoned condition is of the order of 0·64 (40 lb./ft.^{3}), about the same as teak.

Seasoning and movement. The sawn timber dries fairly slowly, with the minimum of distortion and splitting. The shrinkage in drying is very low and, once dried, muninga is extremely stable in use, comparable to teak.

Strength and bending properties. Muninga is roughly comparable to iroko in strength and, like iroko, is only moderately good for steam bending.

Durability and preservative treatment. It is known to be resistant to fungal decay and insect attack, including termites, and is classed as durable. The heartwood is not amenable to impregnation; however, muninga is not a timber which needs preservative treatment.

Working and finishing properties. The timber can be worked readily with all hand and machine tools, with only a moderate blunting effect. Straight-grained material finishes cleanly and smoothly but in planing figured wood it is advisable to reduce the cutting angle to about 20° to minimise tearing. The wood turns well, takes nails, screws and glues readily and gives excellent results with conventional finishing treatments.

Uses. Muninga combines a handsome, dignified appearance with excellent working and finishing qualities, durability and stability in service. It is eminently suitable for high-class furniture, cabinet work and joinery, including exterior doors, and for decorative flooring with subfloor heating. In East Africa it is also used for shipbuilding and for plywood manufacture.

Muninga

Flat-cut

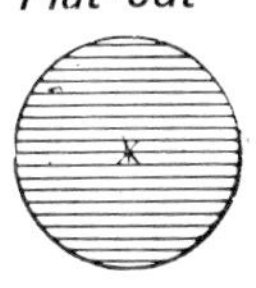

Reproduced two-thirds actual size

KO

Mutenye

[*Guibourtia arnoldiana*]

The Congo name mutenye has been adopted by the timber trade. Other local names are benge and libengi.

Distribution and supplies. Mutenye is a comparatively small tree, rarely exceeding 27 m. (90 ft.) in height, with an irregular bole, 9–18 m. (30–60 ft.) long and 400–750 mm. (say, 15–30 in.) in diameter. It is not common and has a somewhat restricted distribution, in the westernmost part of the former French and Belgian Congos and in the Portuguese enclave of Cabinda.

General description. The heartwood is pale yellowish-brown to medium-brown, sometimes with a faint reddish tinge, with grey or almost black veining. In appearance it is very like the closely related ovangkol (see p. 162) but a little finer textured. It is moderately hard and heavy, about the same weight as bubinga, another allied species of *Guibourtia* (see p. 98). Mutenye has an interlocked and sometimes wavy grain, which, combined with its natural colour variation, gives the wood a highly decorative appearance.

Technical properties. Little is known about the technical properties of mutenye. Its general drying characteristics are uncertain but it is reported to have a high shrinkage although, once dry, limited tests suggest that it has medium movement in service. It is fairly dense but non-siliceous, and can be sawn satisfactorily if rather slowly; planing presents some difficulty, mainly with quartered stock, because of the character of the grain but, when surfaced, it can be given a satisfactory finish. It can be sliced to produce a high-quality veneer. Limited tests suggest that its steam-bending characteristics, although not good, are somewhat better than those of most tropical hardwoods.

Uses. Mutenye is an attractive timber, occasionally marketed in the form of lumber but mainly as veneer. It is a useful addition to the range of walnut-like woods and, as veneer, is used for cabinet work, furniture and interior decorative work; it is too heavy to find extensive use as solid timber but is suitable for turnery, flooring and some furniture parts.

Mutenye

Quarter-cut

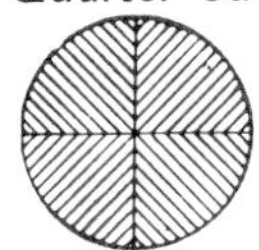

Reproduced actual size

Niangon

[*Tarrietia utilis*]

The Ivory Coast name niangon has been adopted as the international trade name for this timber. The Ghana variant is nyankom.

Distribution and supplies. *Tarrietia utilis* is reported to be common in the West African coastal forests from Sierra Leone, through Liberia and the Ivory Coast, to Ghana. A closely allied species, *T. densiflora*, occurs in Gaboon; the timber is said to be similar to niangon and is shipped under the same name. Niangon is a tree of moderate size with a bole length of 18 m. (60 ft.), or so, and an average diameter of 0·6–0·9 m. (2–3 ft.). The timber has been shipped to France from the Ivory Coast and Gaboon in considerable quantities since the nineteen thirties. More recently it has been exploited for export in Ghana and is now available in Britain as square-edged lumber, 25–50 mm (1–2 in.) thick, 150 mm. (6 in.), and up, wide, in lengths of 2 m. ($6\frac{1}{2}$ ft.) and up.

General description. Niangon resembles a rather open-grained mahogany but is somewhat heavier, average density about 0·64 (40 lb./ft.^{3}), and has a resinous content which makes the surface greasy or sticky to the touch, suggesting teak. The grain is commonly interlocked, producing on quarter-sawn stock an irregular stripe figure enhanced by a rather unusual fleck due to the large rays.

Seasoning and movement. It dries fairly rapidly, with slight distortion, end splitting and surface checking.Dimensional movement in service is classed as medium.

Strength and bending properties. Niangon is as strong as mahogany in most respects but appreciably harder and more resistant to shear, splitting and compression forces. It is classed as moderately good for steam bending.

Durability and preservative treatment. The heartwood is classed as moderately durable, and has been found suitable for exterior joinery. It is extremely resistant to preservative treatment.

Working and finishing properties. It works fairly easily with hand and machine tools, and has comparatively little blunting effect. Quarter-sawn stock tends to tear in planing; the cutting angle should be reduced to 15°. It tends to split when nailed. Difficulties in finishing, due to the resinous nature of the wood, can usually be overcome by a preliminary alkaline wash, e.g. with a solution of ammonia. It paints and polishes well provided that the grain is filled.

Uses. Niangon has been widely used on the Continent for exterior and interior joinery with a natural finish, also in ship and boat building for hulls, deck-houses and cabin fittings, for greenhouse construction and flooring. In Britain it has been used to a limited extent for furniture interiors. It is considered a useful addition to the group of mahogany-type timbers, especially as it may be obtainable at a lower price than true mahogany.

Niangon

Flat-cut

Reproduced actual size

Obeche or Wawa

[*Triplochiton scleroxylon*]

The Nigerian name obeche and the Ghana name wawa have been adopted as alternative British Standard names for the timber of this species; obeche is the usual trade name in Britain. The names ayous and samba refer to timber originating in Cameroon and the Ivory Coast, respectively.

Distribution and supplies. Obeche is widely distributed in West Africa; it is one of the commonest timber trees of Nigeria, Ghana and the Ivory Coast, which are the principal sources of supply. In terms of quantity exported, obeche is the most important West African timber after gaboon (okoumé). It is readily obtainable in large sizes as logs 0·5–1·8 m. (say, $1\frac{1}{2}$–6 ft.) in diameter, mostly 0·6 m. (2 ft.), and up, and 3·5–7·5 m. (12–25 ft.) long, and as sawn timber 19–100 mm. ($\frac{3}{4}$–4 in.) thick and 150–550 mm. (6–22 in.) wide, in lengths of 2–5·5 m. ($6\frac{1}{2}$–18 ft.). Special sizes can be obtained to order. It is also available as plywood.

General description. The wood is creamy-white to pale-straw in colour with no clear distinction between sapwood and heartwood, though the wide sapwood is more susceptible to discoloration and insect attack. It is the lightest low-cost utility hardwood in general use, the density being about 0·38 (24 lb/ft.^{3}) seasoned. The grain is slightly interlocked; the texture open. When cut on the quarter and stained it has some resemblance to African mahogany. Large logs commonly contain brittleheart.

Seasoning and movement. The timber can be dried fairly rapidly, with little degrade other than a tendency to distort. Rapid drying is advisable to reduce the risk of fungal stain. The movement after seasoning is rated as small.

Strength and bending properties. It is fairly elastic and resilient, considering its weight, but should not be used for purposes where strength is critical. Wood from the centre of large logs is inclined to be brittle (brittleheart). On the basis of laboratory tests it is classed as moderately good for steam bending.

Durability and preservative treatment. Obeche is not resistant to decay or staining fungi. Freshly felled logs are extremely prone to attack by pinhole borer beetles, and seasoned timber is often infested by powder-post beetles. In regions where termites are present obeche is very liable to be damaged. The heartwood resists preservative treatment.

Working and finishing properties. Although it cannot be described as hard, the wood is firm under the tool, and even in texture. It works very easily with hand and machine tools, and does not blunt cutting edges of tools very quickly. In end-grain working, the timber may show a tendency to crumble, unless the tools are kept sharp, and edges are not allowed to become thick. It can be turned but is rather soft for this type of use as centres are apt to sink in. For jointed work, gluing is preferable to nailing or screwing, except for very light work. It stains and polishes well (the grain needs to be filled). The wood takes paint well with normal primers.

Uses. Obeche is readily obtainable in large sizes, clear of defects, and at a fairly low price and is ideal for mass-production work. It is used in the manufacture of lower-priced domestic cabinet work and kitchen furniture, and for interior joinery and similar purposes where American whitewood or joinery-grade softwood were formerly specified; also for boxes and packing cases where a good appearance is required, since less wastage occurs in conversion. Obeche should never be used without preservative treatment in exposed or damp situations.

Obeche

Flat-cut

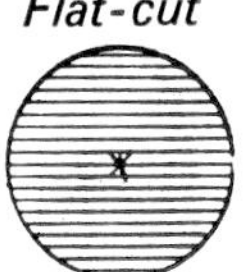

Reproduced two-thirds actual size

Ogea

[*Daniellia ogea*]

Ogea or oziya is the Nigerian name; other trade names are faro (Spanish Guinea) and incenso (Portuguese Guinea). *Daniellia ogea* appears to be the most important of the numerous species of *Daniellia* from the point of view of timber production for the export market. Other species are used locally.

Distribution and supplies. Ogea is reported to be very common in the rain forest of Southern Nigeria where it grows to large dimensions with an exceptionally long, clean, cylindrical bole up to 30 m. (100 ft.) or more, in length, yielding logs 0·7–2 m. ($2\frac{1}{2}$–$6\frac{1}{2}$ ft.) in diameter. It is also found in other parts of West Africa. Logs are exported to Europe in relatively small quantities, mainly for plywood manufacture. Decorative veneers have been imported into Britain from the Continent.

General description. A light-weight, utility hardwood, density varying from 0·42–0·58 (6–36 lb./ft.^{3}), seasoned; pinkish to reddish-brown, marked with darker streaks, with an exceptionally wide sapwood, 100–180 mm (say, 4–7 in.), likely to become discoloured if conversion is delayed. The heartwood is apt to be gummy and is appreciably denser than the sapwood. The grain is more or less interlocked, the texture rather coarse. Cross-breaks are liable to occur towards the centre of large logs.

Seasoning and movement. Generally the sawn timber can be dried rapidly, with no serious degrade. The movement in service is classed as medium.

Strength and bending properties. In strength, ogea is comparable to Baltic redwood but is appreciably harder and more resistant to splitting. It is almost useless for steam bending.

Durability and preservative treatment. Both heartwood and sapwood are perishable; the wide sapwood is particularly susceptible to insect attack and sap stain. The sapwood is permeable to preservative treatment, the heartwood is resistant.

Working and finishing properties. The timber is fairly easy to work but quarter-sawn material tends to tear, and the surface is apt to be rough and woolly unless tools are kept sharp. It nails well, takes stain readily but needs careful filling before polishing. The finishing of gummy material is facilitated if it is wiped over with a spirit solvent. It can be glued satisfactorily.

Uses. In the form of sawn timber ogea is classed as a relatively low-grade, light hardwood, suitable in place of softwood for rough construction, boarding, boxes and packing cases and similar purposes where strength, durability and appearance are unimportant. By reason of the large size and excellent form of the logs it has been found suitable in Europe for the manufacture of plywood (mainly for core stock). Selected logs yield a decorative veneer.

Ogea

Flat-cut

Reproduced actual size

Okan

[*Cylicodiscus gabunensis*]

Distribution and supplies. Widely distributed in the rain forests of West Africa, and particularly common in Ghana, where it is known as denya, and in Nigeria. A large tree with a well-shaped bole clear of branches to a height of 24 m. (80 ft.). The average diameter is just over 1 m. (say, 3–4 ft.). The timber is not regularly exported in quantity but can be supplied to order in the dimensions required.

General description. Very hard and heavy, average density about 0·96 (60 lb./ft.^{3}) in the seasoned condition, i.e. about the same density as greenheart. The sapwood is fairly wide, 50–75 mm. (2–3 in.), the heartwood a striking greenish-brown colour, darkening to reddish-brown. The grain is strongly interlocked, the texture rather coarse.

Seasoning. It dries slowly, with a marked tendency to split and check.

Strength and bending properties. Limited tests indicate that okan is similar in strength to the Australian timber karri. It is not amenable to steam bending.

Durability. Okan is resistant to fungi, termites and marine borers, and is placed in the highest durability class with such timbers as greenheart, ekki, jarrah and teak.

Working and finishing properties. Being exceptionally hard and heavy, with interlocked grain, it is naturally difficult to cut; sawmill conversion is practicable, however. It should be pre-bored for nailing.

Uses. Okan is essentially a heavy constructional timber for use under exposed conditions, notably for marine construction, as piling, wharf decking, etc. It is also suitable for heavy-duty flooring in warehouses and factories.

Okan

Flat-cut

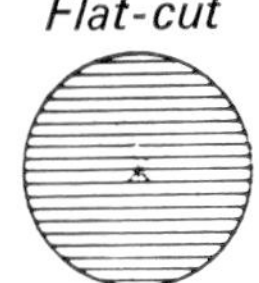

Reproduced actual size

Okwen

[*Brachystegia nigerica* and *B. kennedyi*]

The name okwen is applied to two closely allied species of *Brachystegia* in Nigeria, as indicated above. There are many other species, widely distributed in tropical Africa.

Distribution and supplies. Okwen is of very common occurrence in Southern Nigeria. The trees are of large dimensions, averaging nearly 1·25 m. (4 ft.) in diameter. Because of the large proportion of sapwood and the tendency to shake, logs are normally converted to quarter-sawn, square-edged stock, 75–250 mm. (3–10 in.) wide. Selected logs are sliced for veneer.

General description. A timber of medium density, suggesting a low-grade mahogany in general appearance but somewhat darker and heavier, about the same weight as oak. The grain is heavily interlocked, producing a pronounced stripe or roe figure on quarter-sawn timber and sliced veneer.

Technical properties. The timber seasons slowly, with some distortion. Dimensional movement in service is classed as medium. Strength properties are similar to those of oak. The heartwood is only moderately durable and is extremely resistant to impregnation. Okwen is rather hard to work and has a fairly severe blunting effect on cutting tools, especially saw teeth. The interlocked grain makes a smooth surface finish difficult to produce.

Uses. Being cheap, plentiful and strong for its weight, okwen is suitable for constructional work and rough boarding in situations where a high degree of resistance to decay and insect attack is not required. Otherwise it appears to have little to commend it for the export market, except perhaps for the manufacture of decorative veneer and for flooring.

Okwen

Flat-cut

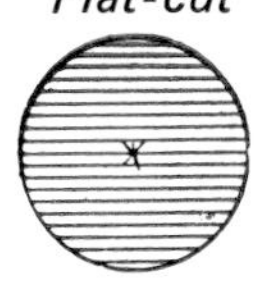

Reproduced actual size

Olive, East African

[*Olea hochstetteri*]

Distribution and supplies. Compared with the European olive of the Mediterranean region, which is a small, stunted tree of poor shape, the East African olive is a large tree growing to a height of 27·5 m. (90 ft.) and a diameter of nearly 1 m. (say, 3 ft.). The average specimen has a clear bole of 6–7·5 m. (20–25 ft.) and a diameter of 0·6 m. (2 ft.). It is common in Kenya, which is the main source of supply, and also occurs in Tanzania, Uganda, Ethiopia, Sudan and the highlands of the Congo region. The timber is exported mainly in the form of flooring strips.

General description. A very hard, heavy wood, density about 0·88 (55 lb./ft.^{3}) in the seasoned condition, of fine, even texture with a characteristic decorative appearance, light-brown marked with irregular, dark greyish-brown veins and streaks.

Seasoning and movement. It dries slowly, with a tendency to check and split. The dimensional movement in service is classed as large.

Strength and bending properties. The strength properties are outstanding. In the seasoned condition it is as strong, or stronger, than Canadian hard maple, although it has a lower resistance to splitting. The steam-bending properties are rather variable; on average it is classed as moderate.

Durability. It is believed to be resistant to fungal and insect attack, including termites.

Working and finishing properties. Being exceptionally hard and dense, it is difficult to work but the blunting effect is not excessive. Provided that sharp tools are used it takes an excellent finish and is particularly good for turnery.

Uses. Outside the countries of origin it is used principally for flooring. It is highly resistant to wear and is very suitable for all types of heavy-duty floor as an alternative to hard maple; also for decorative flooring, as in public buildings and ballrooms. Its attractive appearance, combined with good working and finishing properties, make it suitable for various kinds of turnery and ornamental work.

Other species of interest. *A closely allied species is loliondo or Elgon olive* (Olea welwitschii), *also from East Africa. It is available in the form of flooring strips and blocks and, like East African olive, is suitable for heavy-duty pedestrian and industrial traffic. It is not so decorative as East African olive.*

Olive

Flat-cut

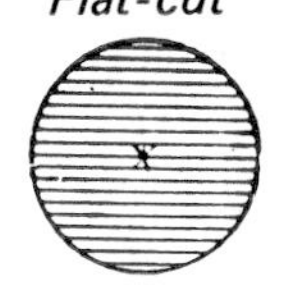

Reproduced actual size

Opepe

[*Nauclea diderrichii* formerly called *Sarcocephalus diderrichii*]

The Nigerian name opepe has been adopted as the British Standard name. The timber is also known as kusia (Ghana), badi (Ivory Coast) and bilinga (Cameroon) according to the country of origin.

Distribution and supplies. The tree is widely distributed in West Africa, from Guinea to the Congo, and the timber is shipped in large or small quantities from most of the timber-producing countries, both in log form and as sawn timber. It is a large, unbuttressed tree with a long, clean bole up to 24 m. (80 ft.), or more, long, yielding logs for export commonly 1–1·25 m. (say, 3–4 ft.) in diameter, 3·5–9 m. (12–30 ft.) long. Square-edged timber is generally available in thicknesses of 25–150 mm. (1-6 in.), widths of 150–450 mm. (6–18 in.), in lengths of 2–6 m. ($6\frac{1}{2}$–20 ft.), and as flooring strips.

General description. The heartwood is a striking yellow or orange-brown, giving the wood a distinctive appearance, often enhanced on quartered surfaces by a stripe or roe figure due to the interlocked or wavy grain. The general direction of the grain is sometimes straight but, more often, somewhat irregular, and occasionally markedly so; the texture is fairly coarse. Opepe is of the same order of density as European oak, about 0·74 (46 lb./ft.^{3}), seasoned.

Seasoning and movement. Quarter-sawn material seasons fairly quickly, with little degrade, but flat-sawn timber is apt to check and split, and serious distortion occasionally develops. In thick sizes the timber dries rather slowly; once dry it has only a small movement with changes in relative humidity.

Strength and bending properties. Opepe, when seasoned, is superior to oak in compression along the grain, in bending and stiffness and in hardness on the side grain; it is somewhat inferior in its resistance to splitting and ability to withstand suddenly applied loads. Limited tests indicate that it is not very suitable for steam bending.

Durability and preservative treatment. Heartwood is noted for its high resistance to fungal attack, termites and marine borers; it is moderately resistant to preservative treatment but the sapwood is permeable.

Working and finishing properties. Opepe presents little difficulty in working and, although somewhat variable according to density, its dulling effect on cutting edges is generally slight. Flat-sawn material planes to a smooth finish but quarter-sawn timber with interlocked or wavy grain tends to pick up, and a 10° cutting angle is necessary for a satisfactory finish. The timber takes screws fairly well but tends to split when nailed. Coarse textured, it requires filling to polish effectively.

Uses. A strong, stable and durable timber, particularly suitable for use in large sizes for structural purposes such as piling wharf and jetty decking and other dock and marine work; in small dimensions care is necessary to exclude timber with markedly irregular or cross grain. Opepe has been used in place of oak in boat building, and is suitable for exterior joinery provided that some surface checking is tolerated; it makes an attractive flooring for normal conditions of pedestrian traffic. Its vivid colour and striking figure on quartered surfaces have decorative possibilities.

Opepe

Flat-cut

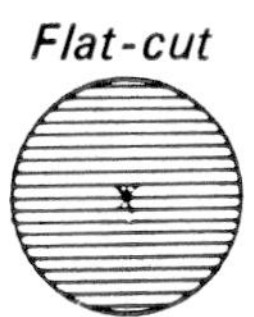

Reproduced two-thirds actual size

Ovangkol or Ehie

[*Guibourtia ehie*]

Distribution and supplies. This species occurs in the Ivory Coast, Ghana, Southern Nigeria and Gaboon. It is a moderately tall tree up to 30 m. (100 ft.) in height and 0·9 m. (3 ft.) in diameter above the buttresses. Little is known about the supplies available; logs have been imported in small quantities for cutting into decorative veneer.

General description. The heartwood is yellow-brown to chocolate-coloured with grey to near-black stripes; when fresh it has a strong smell but this disappears on drying. It is similar in appearance to Queensland walnut but in the natural state is paler in colour and, like the Australian timber, differs from European walnut in its more regular striping. The timber is slightly coarser-textured and probably a little heavier than Queensland walnut; the grain is interlocked.

Technical properties. Little is known of its technical properties. It is a fairly dense, although non-siliceous wood and may present a moderate resistance in sawing and machining, with care necessary in finishing quartered stock because of interlocked grain.

Uses. Ovangkol has been used to a limited extent as decorative veneer. It should provide a useful addition to the range of walnut-like woods, and as solid timber appears to be worth consideration for cabinet work, high-grade furniture, interior decorative work, turnery, flooring, etc.

Other species of interest. *Allied species of* Guibourtia *furnish decorative timbers somewhat similar to ovangkol, notably bubinga* (*see p. 98*), *mutenye* (*see p. 146*) *and Rhodesian copalwood* (G. coleosperma).

Ovangkol/Ehie

Quarter-cut

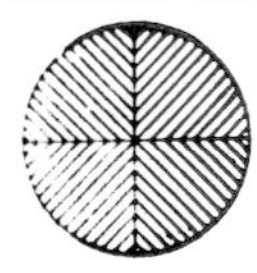

Flat-cut

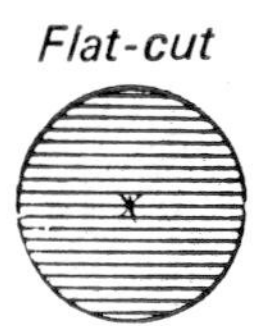

Reproduced actual size

Pterygota

[*Pterygota bequaertii* and *P. macrocarpa*]

The botanical name *Pterygota* has been proposed for the timber of these two closely allied species, which has also been marketed under the Ghana name awari or ware. In Nigeria it is known as kefe, and in the Ivory Coast as koto.

Distribution and supplies. Both species are large trees common in the rain forests of West Africa. The bole is straight and cylindrical, 0·5–1·25 m. (say, 1½–4 ft.) in diameter. As yet, the timber has been exported in small quantities only but supplies could be increased to meet further demand.

General description. Pterygota is a uniformly pale, straw-coloured to almost white, utility hardwood, coarse-textured with a shallowly interlocked grain. When flat-sawn it has a plain appearance with a faint growth-ring figure but on accurately quartered surfaces there is a conspicuous fleck due to the high rays. The timber of *P. macrocarpa* is paler and at 0·56 (35 lb./ft.^{3}), somewhat lighter in weight than that of *P. bequaertii*, which averages about 0·66 (41 lb./ft.^{3}) when dry. However, both species are somewhat variable in colour and weight, making their separation somewhat doubtful. The timber has an unpleasant smell when green but this disappears on drying.

Seasoning and movement. Pterygota dries fairly rapidly, with only a slight tendency to surface splitting and, in general, little tendency to distort. Once dry, it is rated as having medium movement in service.

Strength and bending properties. The strength of pterygota is somewhat variable according to its weight. Timber of *P. bequaertii*, compared with European beech, is on average a little lighter in weight, slightly inferior in bending strength and stiffness but slightly more resistant to crushing forces. Limited tests indicate that it has very poor bending properties. *P. macrocarpa* is rather less strong, being 30 to 40 per cent weaker than beech in bending, stiffness and compression.

Durability and preservative treatment. Pterygota is particularly susceptible to deterioration by staining fungi and pinhole borers unless rapidly extracted, converted and given a combined fungicidal and insecticidal dip. When dry, the timber is rated non-durable under conditions favouring fungal attack, and the sapwood is susceptible to powder-post beetles. It is readily treated with preservatives.

Working and finishing properties. Pterygota cuts fairly easily and works well with machine and hand tools in a sharp condition. There is some tendency to a fibrous finish, and to avoid tearing when machining quartered stock with interlocked grain a cutting angle of 20° is recommended. It nails well and can be glued satisfactorily. The timber can be rotary peeled and sliced for decorative veneer.

Uses. Pterygota requires rapid extraction and conversion to be marketed in clean condition but when free from stain and insect damage it is an attractive, pale-coloured if fairly plain wood. It has been suggested as an alternative to obeche but is considerably harder and heavier, more in the class of ramin and beech, although somewhat coarser textured and with a shallowly interlocked grain compared with the more usual straight grain of these timbers. It is suitable for furniture manufacture, interior joinery and carpentry and as plywood.

Pterygota

Quarter-cut

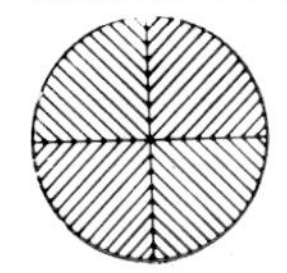

Reproduced actual size

'Rhodesian Teak'

[*Baikiaea plurijuga*]

This timber is called Rhodesian, or Zambian teak because of its outstanding strength, durability and stability. In Rhodesia and Zambia (formerly Northern Rhodesia) it is regarded as the local equivalent of teak.

Distribution and supplies. This species is limited to the relatively dry region drained by the Upper Zambesi and Okavango rivers, an area including the western parts of Zambia and Rhodesia and adjoining portions of Angola, South-West Africa and Botswana. Compared with the giants of the tropical African rain forests it is a small tree, commonly 9–15 m. (30–50 ft.) high with a clear bole of 3–4·5 m. (10–15 ft.), up to 0·75 m. ($2\frac{1}{2}$ ft.) in diameter. It is the most important indigenous timber tree exploited commercially in Zambia and Rhodesia where it is largely used for railway sleepers, construction work and furniture. The export trade is practically limited to flooring blocks manufactured from offcuts.

General description. A handsome wood of distinctive appearance. The surface of freshly manufactured timber is light-brown marked with irregular black lines or flecks. On exposure to light it soon changes to reddish-brown, and after some years the red component disappears, leaving a beautiful dark-brown colour. The texture is fine and even, giving a smooth, hard surface; the grain is straight or slightly interlocked. The average density is about 0·9 (56 lb/ft.^{3}), i.e. appreciably harder and heavier than true teak.

Seasoning and movement. 'Rhodesian teak' dries slowly, but with the minimum of degrade. It is dimensionally stable under varying conditions of atmospheric humidity, being similar to mahogany in this respect, but less stable than true teak.

Strength and bending properties. Laboratory tests on material in the form of flooring blocks have shown that it is appreciably harder than Canadian hard maple. In steam-bending tests it tended to buckle and was rated as only moderately good.

Durability and preservative treatment. It is well known for its outstanding durability under tropical conditions, and is resistant to termites. The heartwood is extremely resistant to impregnation with preservatives.

Working and finishing properties. The timber is difficult to work and has a fairly severe blunting effect on cutting edges. A clean finish is usually obtained in planing. It should be pre-bored for nails, to prevent splitting.

Uses. Outside the country of origin 'Rhodesian teak' is mainly used for flooring. It makes a highly decorative, hard-wearing floor, wears smoothly under all conditions of pedestrian traffic, and is very stable under varying conditions of temperature and humidity.

'Rhodesian Teak'

Quarter-cut

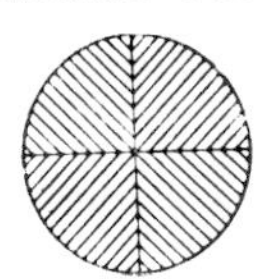

Reproduced actual size

Sapele

[*Entandrophragma cylindricum*]

The trade name sapele mahogany or sapele wood was originally applied to consignments of 'mahogany' shipped from the Nigerian port of Sapele towards the end of the last century. It is now restricted to the timber of one botanical species, *Entandrophragma cylindricum*, more or less irrespective of origin. The French equivalent is sapelli or acajou sapelli.

Distribution and supplies. The tree is widely distributed in West and Central Africa, extending eastwards to Uganda. In the Ivory Coast, Ghana and Nigeria it occurs abundantly and is one of the principal species of timber exported; it is also shipped from Cameroon, Congo and Gaboon. It grows to a large size with a straight, cylindrical bole clear of branches to a height of 25–30 m (80–100 ft.). The timber is readily obtainable as logs 0·6–1·25 m. (2–4 ft.) or more, in diameter and 3·5–9 m. (12–30 ft.) in length and as sawn timber 150–600 mm. (6–24 in.) wide, 25–100 mm. (1–4 in.) thick, in lengths of 2–6 m. (6½–20 ft.), or larger; also as flooring strips, furniture squares, interior and exterior grade plywood and decorative veneers.

General description. Sapele is appreciably harder and heavier and has a finer texture than African mahogany; the average density is about 0·64 (40 lb./ft ^{3}), seasoned. The colour is generally a fairly dark reddish or purplish-brown. It has a natural lustre and a pleasant, cedar-like scent when freshly cut. The straight stripe or roe figure of quarter-cut material, due to the regular interlocked grain, is a well-known characteristic of sapele. A variation of this figure is the broken roe or mottle. Large logs are liable to ring- or cup-shakes but the timber is generally free from the cross-breaks which occur in African mahogany.

Seasoning and movement. Sapele seasons slowly, with a marked tendency to distort; kiln-dried stock may therefore need reconditioning. It is less stable than African mahogany and should preferably be cut on the quarter.

Strength and bending properties. It is in the same strength class as oak, i.e. it is considerably stronger than either African or American mahogany. It is unsuitable for steam bending.

Durability and preservative treatment. The timber is not readily attacked by fungi or insects but is said to be susceptible to pinhole borer damage in the round. It is resistant to preservative treatment.

Working and finishing properties. Although it is somewhat harsher than African mahogany and takes the edge off cutting tools rather more quickly, it is not a difficult wood to work, except that in planing and moulding the surface is likely to tear up owing to the interlocked grain; this trouble may be largely overcome by reduction of the cutting angle to 15°. Glued joints are usually good and sound. It takes polish satisfactorily and can be painted if required, using normal primers.

Uses. Sapele is well known as a decorative sliced veneer for high-class furniture, piano cases and panelling. It is also used for interior parts of furniture such as drawer sides and rails and runners. It makes a strong plywood. Its excellent strength and good appearance are recognised in its use for joinery items of a substantial nature, e.g. staircases, heavy window frames and flooring. It is commonly selected for the framework of all types of road vehicles.

Other species of interest. Entandrophragma candollei, *known as omu or kosipo, is dull-brown or purplish-brown with fairly coarse texture. In its technical properties it is similar to sapele and can be used in the same way.*

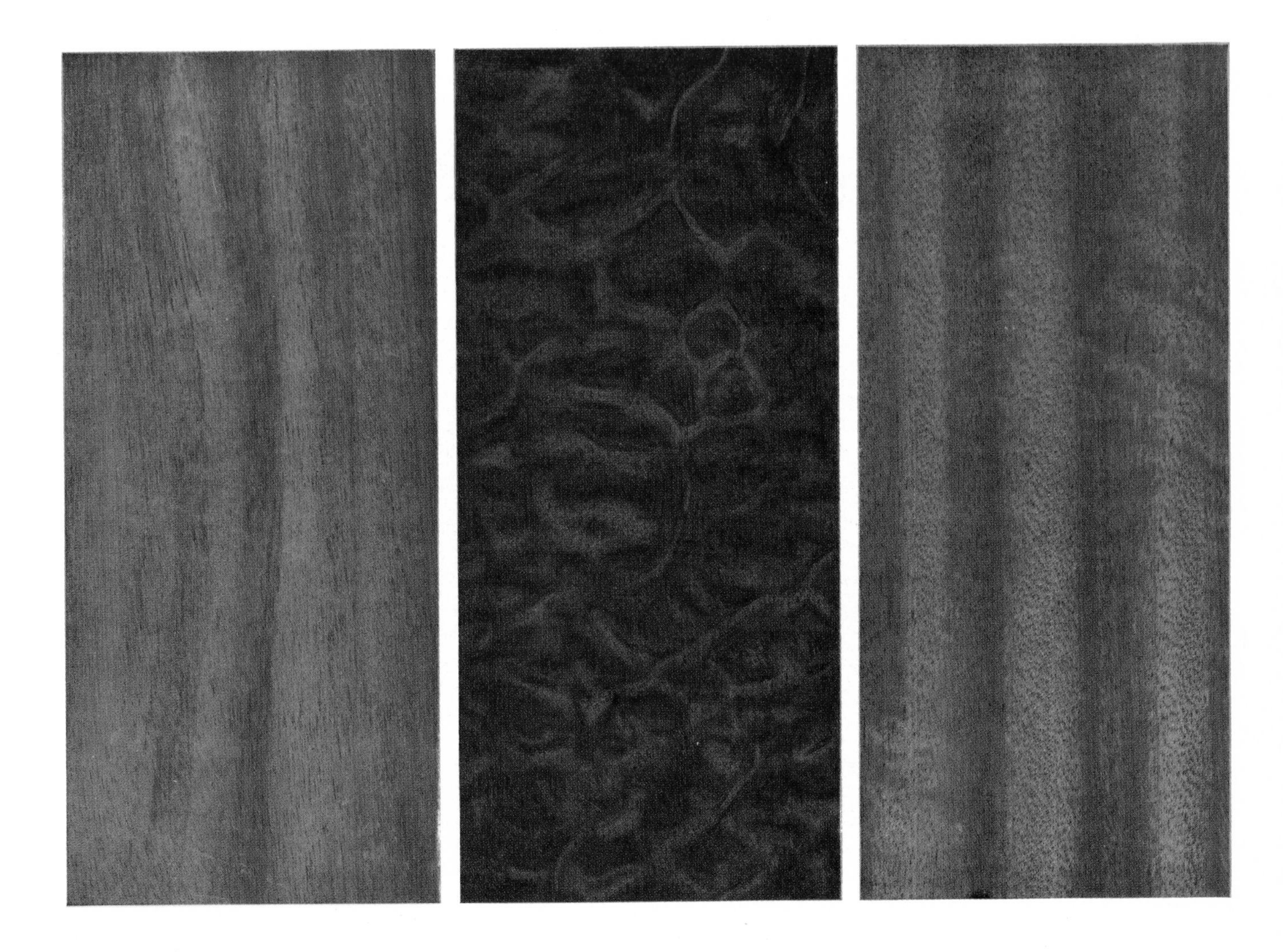

Sapele

Flat-cut

Burr

Quarter-cut

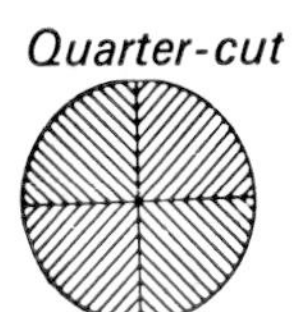

Reproduced actual size

Tchitola

[*Oxystigma oxyphyllum* formerly called *Pterygopodium oxyphyllum*]

The Congo name, tchitola, appears to be the principal trade name for this timber on the international market. It is also known as tola mafuta, tola chinfuta or, simply, tola (Angola), lolagbola (Nigeria) and tola walnut (Britain). Note that the name tola is also used in the Portuguese African territories for the timber more familiar in Britain as agba (see p. 86). Tchitola is easily distinguished from agba by the colour of the heartwood.

Distribution and supplies. The principal source of supply is the Congo region, including Angola. The timber is available in Nigeria, Cameroon and Gaboon also. A large tree of good shape yielding cylindrical logs 0·6–0·9 m. (2–3 ft.) in diameter. The timber is exported both in log form and as veneer.

General description. Tchitola is of interest mainly for the decorative character of the heartwood which is brown with darker markings, suggesting walnut. It is a gummy wood but the gum is mainly concentrated in the fairly wide sapwood zone. It is of medium density, about 0·6 (38 lb./ft.^{3}) in the seasoned condition, i.e. somewhat lighter than walnut. The texture is moderately coarse, the grain straight or slightly interlocked.

Seasoning and movement. It is said to air dry with very little distortion or splitting. Once dry it is reported to be stable in use under changing conditions of relative humidity.

Strength properties. The limited data suggest that tchitola is above average in strength for its weight. It compares closely with sapele in bending strength, stiffness, compression and resistance to impact but is weaker in shear, and less hard.

Durability and preservative treatment. Laboratory tests indicate that the heartwood is variable in durability, some timber having a moderate resistance to fungal attack but occasional pieces giving a less satisfactory performance. The sapwood is perishable and susceptible to powder-post beetle attack. It is uncertain whether the timber can be effectively treated with preservative.

Working and finishing properties. Tchitola gives little trouble in sawing and planing, apart from a tendency for clogging of tools by gum. It machines to a smooth surface and can be finished satisfactorily with—in general—little trouble from gum. It is said to nail and screw easily and to hold both well. Tchitola can be rotary peeled successfully and glues well.

Uses. On the European market tchitola is known mainly as a decorative veneer for television cabinets, for example. In South Africa it is used for the manufacture of general-purpose plywood.

Tchitola

Quarter-cut

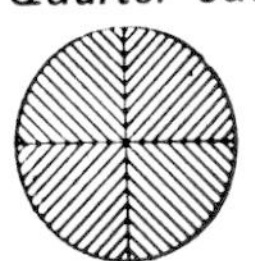

Reproduced actual size

Utile, Sipo or Assié

[*Entandrophragma utile*]

The trade name utile, taken from the botanical name, is commonly used in Britain. The Ivory Coast name sipo or acajou sipo is in general use on the European market. Timber from Cameroon is sometimes known by the local name of assié or acajou assié.

Distribution and supplies. The tree has a wide distribution in West and Central Africa eastwards to Uganda. The timber is exported—mainly in log form, though the proportion of sawn timber is increasing—from all the timber-producing countries of West Africa, particularly the Ivory Coast and Ghana where it is one of the principal commercial species. It is a fine tree with a straight, cylindrical bole clear of branches to a height of 21–24 m. (70–80 ft.). Logs are 0·6–1·8 m. (2–6 ft.) in diameter, 3·5–9 m. (12–30 ft.) long. Sawn timber is generally obtainable in widths up to 600 mm. (24 in.), thicknesses of 16–100 mm. ($\frac{5}{8}$–4 in.) and lengths of 2–6 m. (6$\frac{1}{2}$–20 ft.), or in larger sizes to order. Utile is also available as plywood and, occasionally, as decorative veneer.

General description. A timber of the mahogany type, similar to sapele in appearance and general character but more open in texture; also the interlocked grain is less regular, so the stripe figure characteristic of sapele is usually much less pronounced. The heartwood is a fairly uniform reddish-brown, density about 0·66 (41 lb./ft.^{3}), seasoned, and lacks the cedar-like scent characteristic of sapele. Compared with the general run of African mahogany, utile is darker in colour, harder and heavier by some 5 or 6 lb./ft.^{3}.

Seasoning and movement. There is a tendency, in some parcels, for distortion to occur, especially in boards with very irregular grain. The warping is most severe in flat-sawn stock but even so is on average less pronounced than in sapele. It can be reduced to a minimum by quartering and then piling out of direct sunlight or strong air currents. The timber may be dried satisfactorily in a kiln provided that a mild drying schedule is used. Dimensional movement in service is classed as medium; it has the reputation of being somewhat better than sapele in this respect.

Strength and bending properties. Utile is slightly inferior to sapele in strength but is considerably stronger than African mahogany. It is unsuitable for steam bending.

Durability and preservative treatment. The wood is not very readily attacked by wood-destroying fungi but is sometimes damaged by pinhole borers. It is extremely resistant to preservative treatment.

Working and finishing properties. Some logs of utile have proved rather difficult to convert, the saw teeth rapidly becoming blunted. This is particularly the case with circular saws when the pitch is too low, but is much less frequent if the teeth are widely spaced. Generally, utile gives less trouble in planing than sapele as the grain is less interlocked. For stock having wild grain the cutting angle should be reduced to 20° or even less, but this involves greater power consumption. In other respects the timber resembles sapele.

Uses. Utile is largely used for furniture interiors in the same way as sapele; the two species are sometimes mixed and marketed as sapele/utile. It is employed fairly widely in the construction of road vehicles and for interior work in railway coaches and ships; also for the manufacture of utility plywood. Selected logs are sliced for decorative veneer.

Utile/Sipo/Assié

Quarter-cut

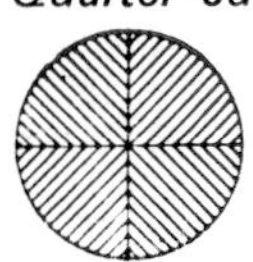

Reproduced actual size

Wenge

[*Millettia laurentii*]

Distribution and supplies. A small tree of the Congo region, which rarely exceeds 0·6 m. (2 ft.) in diameter. The timber is exported in the log and as lumber, and can also be obtained in the form of veneer.

General description. The general impression is dark-brown but close examination shows the wood to comprise alternate layers of dark and light-coloured tissue, giving a distinctive decorative appearance to flat-sawn timber and tangentially cut veneer. It is hard and dense, of the order of 0·80 (50 lb./ft.^{3}), fairly coarse textured and generally straight grained.

Seasoning and movement. It is reported to dry rather slowly, but without much distortion. Dimensional movement in service is small.

Strength. Wenge has the reputation of being strong and very elastic with a high resistance to shock; it is said to be similar to hickory in this respect.

Durability. It is believed to be highly resistant to fungal decay and insect attack.

Working and finishing properties. Somewhat hard to work on account of its density and fibrous character. Provided that tools are kept sharp it takes a good natural finish.

Uses. Wenge is more popular on the Continent than in Britain. It is in use there for cabinet making, interior fittings and panelling in ships and railway coaches, for fancy-goods such as brush backs, and for parquet flooring.

Other species of interest. *Panga panga* (Millettia stuhlmannii) *is one of the principal timbers exported from Mozambique, mainly to South Africa, Rhodesia and Europe. It is practically indistinguishable from wenge and is used for the same purposes, being particularly suitable for hard-wearing, decorative flooring.*

Wenge

Quarter-cut

Reproduced actual size

Zebrano

[*Microberlinia brazzavillensis* and *M. sulcata*]

Zebrano is also known as zingana and African zebrawood. Note that the trade name rose zebrano refers to a different timber, berlinia (see p. 94).

Distribution and supplies. Supplies come from Cameroon and Gaboon, in the form of billets and logs up to nearly 1 m. (say, 3 ft.) in diameter, mainly for conversion to veneer, though it is also obtainable in the solid.

General description. A decorative timber of unusually striking appearance, light yellowish-brown with sharply contrasting dark brown or nearly black narrow stripes. The pattern of striping varies considerably from one log to another. The wood has a natural lustre. The density is around 0·64–0·80 (40–50 lb./ft.^{3}) in the seasoned condition.

Technical properties. The timber is difficult to season without distortion and should be quarter sawn for use in the solid. It can be sawn and worked without difficulty by hand and machine tools, finishing to a smooth lustrous surface, and is readily cut into veneers by slicing quartered billets. Veneers need to be carefully handled to prevent cracking.

Uses. Zebrano is used almost exclusively for decorative veneers, usually sliced on the quarter to show the characteristic stripe effect to the best advantage. Solid wood is sometimes used for ornamental turnery and fancy-goods.

Zebrano

Flat-cut

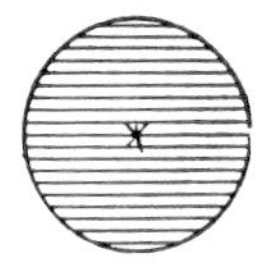

Rotary-cut

Quarter-cut

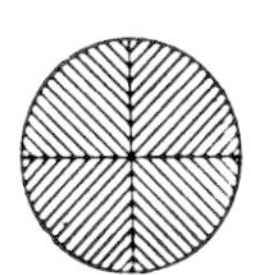

Reproduced actual size

African Timbers
SOFTWOODS

Pencil Cedar, African

[*Juniperus procera*]

Distribution and supplies. African pencil cedar, although widely distributed in East Africa, occurs only at high altitudes and is obtained mainly from Kenya and, to a lesser extent, from Tanzania. The tree reaches 30–37 m. (100–120 ft.) in height and 1·2–1·5 m. (4–5 ft.) in diameter; it has a tapering trunk with a fluted base, and large trees are sometimes hollow. It is exported as lumber and in the form of pencil slats.

General description. African pencil cedar is so called because of the cedar-like scent of the wood and its principal use. It is a medium-weight softwood (conifer) with a fine, even texture and a straight grain. The heartwood is yellowish-brown to reddish-brown, similar in colour and texture to the well-known Virginian pencil cedar (*Juniperus virginiana*) but with less well-defined growth rings. It has an average density of 0·58 (36 lb./ft.^{3}) when seasoned, about 15 per cent higher than Virginian pencil cedar and some 40 per cent higher than incense cedar (*Libocedrus decurrens*), another pencil wood.

Seasoning. A slow-drying timber for which FPRL kiln schedule G has been suggested. It has a marked tendency to fine surface checking on drying, and in thicker sizes is apt to split and check at the ends.

Strength and bending properties. The limited strength data available indicate that African pencil cedar is slightly inferior to Baltic redwood in bending strength, stiffness and compression; it is 20–45 per cent harder than Virginian pencil cedar.

Durability and preservative treatment. The heartwood is rated durable when exposed to fungal attack and is reputed to be moderately resistant to termite attack. It is extremely resistant to preservative impregnation by pressure treatment.

Working and finishing properties. The timber works easily with hand and machine tools, and has little dulling effect on cutting edges. An excellent finish is obtained provided that tools are kept sharp. African pencil cedar whittles cleanly, although it is somewhat harder than Virginian pencil cedar in this respect. It tends to split and requires support when drilled, mortised or end moulded; care is required in nailing and screwing. It can be glued satisfactorily.

Uses. African pencil cedar combines an attractive appearance, durability, ease of working and stability. It largely replaced Virginian pencil cedar for pencils when supplies of American timber became scarce but is now in competition with American incense cedar, which is milder in character and easier to whittle. It is also used to a limited extent for greenhouse construction and furniture manufacture. In East Africa it is used for cabinet work and for joinery, including exterior joinery. It is somewhat soft for flooring. Cedarwood oil is distilled from wood waste.

Pencil Cedar

Quarter-cut

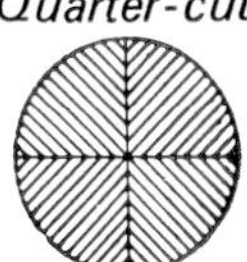

Reproduced actual size

Podo

[various species of *Podocarpus*]

Distribution and supplies. The species of *Podocarpus* which furnish commercial timber are widely distributed in the highlands of East Africa. They attain a height of 30 m. (100 ft.) or more with diameters averaging 0·45–0·65 m. ($1\frac{1}{2}$–$2\frac{1}{2}$ ft.). The timber is shipped as square-edged lumber, 50–300 mm. (2–12 in.) wide, 25–75 mm. (1–3 in.) thick, in lengths up to 5 m. (16 ft.).

General description. The timber of these tropical softwoods (coniferous trees) differs from that of the typical European and North American softwoods in having no clearly defined growth rings and consequently a more uniform texture. It resembles Parana pine in this respect. On average it is similar to Baltic redwood in density, about 0·51 (32 lb./ft.^{3}), seasoned. The colour is usually a uniform light yellowish-brown throughout but streaks of darker-coloured compression wood are sometimes present, and some logs show a small darker-coloured core. The wood is non-resinous and odourless. It is generally straight grained.

Seasoning properties. Podo dries fairly rapidly, with some distortion, checking and splitting. Distortion can be minimised by weighting the stack during seasoning. If compression wood is present considerable longitudinal shrinkage may occur.

Strength and bending properties. Podo is similar to Baltic redwood in strength except that it is appreciably harder. From limited tests it is classed as a moderately good bending wood, i.e. it is probably suitable for solid bends with a radius of curvature from 280–500 mm. (11–20 in.).

Durability and preservative treatment. This timber is classed as non-durable. It is unsuitable for exterior work unless adequately treated with a preservative, but it responds well to impregnation.

Working and finishing properties. The timber is easy to work with hand and machine tools, and being fairly uniform in texture it planes and moulds to a good finish and turns well. It tends to split in nailing unless fine gauge nails are used. It responds well to the usual finishing treatments and can be glued satisfactorily.

Uses. Podo is essentially a joinery-grade softwood. For this class of work it is superior to the general run of European softwoods but care should be taken to eliminate the dark streaks of compression wood which are liable to cause distortion in use. The relatively large sizes obtainable are an advantage for such purposes as shop fitting. It is suitable for the manufacture of fine mouldings because of its good machining and finishing qualities.

Podo

Flat-cut

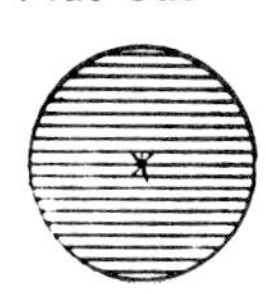

Reproduced actual size

Thuya Burr

[*Tetraclinis articulata*]

Not to be confused with species of the botanical genus *Thuja*.

Distribution and supplies. This little-known tree is confined to dry regions in North-West Africa (Morocco and Algeria) and Malta. It varies in size from a small bush to a medium-sized tree up to 15 m. (50 ft.) high and 300 mm. (12 in.) in diameter. It is known to commerce in the form of burrs which are usually cut into decorative veneers. Supplies are limited and the price is correspondingly high.

General description. The wood is aromatic, yellowish-brown to reddish-brown, and fairly hard and heavy for a softwood. The highly contorted wood of burrs yields beautifully mottled veneers with a figure resembling that of bird's-eye maple or amboyna.

Uses. Thuya burrs have been utilised on a small scale for centuries, mostly in the form of decorative veneer for fine cabinet work. Veneers are still used for cigarette and cigar boxes, presentation caskets, etc. Pieces of solid wood are sometimes made into cigarette cases and other small fancy articles.

Thuya Burr

Reproduced two-thirds actual size

Literature References

General

A Handbook of Hardwoods (Forest Products Research Laboratory, HM Stationery Office, London, 1956)

A Handbook of Softwoods (FPRL, HMSO, 1960)

Nomenclature of Commercial Timbers, including Sources of Supply. British Standards 881 and 589 (British Standards Institution, London, 1955)

Nomenclature Générale des Bois Tropicaux (Association Technique Internationale des Bois Tropicaux, Nogent-sur-Marne, 1965)

Tropical Timber. Statistics on Production and Trade (Organisation for Economic Cooperation and Development, Paris, 1967)

British Timbers. By E. H. B. Boulton and B. A. Jay (A. and C. Black, London, 1946)

Gold Coast Timbers (Takoradi, 1952)

Some Nigerian Woods. By G. von Werndorff and L. Okigbo (Federal Ministry of Information, Lagos. 1964)

A Guide to Building Timbers in Nigeria. By L. Okigbo (Federal Department of Forest Research, Ibadan, 1963)

Catalogue of Kenya Timbers. By S. H. Wimbush (Nairobi, 1950)

Bois Congolais. By L. Lebacq (Mons, 1954)

Seasoning and Movement

Timber Seasoning (Timber Research and Development Association, 1962)

Kiln Operator's Handbook. A Guide to the Kiln Drying of Timber. By W. C. Stevens and G. H. Pratt. (FPRL, HMSO, 1961)

Kiln Drying Schedules. FPRL leaflet No. 42 (HMSO, 1959)

The Treatment of Timber in a Drying Kiln. FPRL leaflet No. 20 (HMSO, 1957)

The Air-Seasoning of Sawn Timber. FPRL leaflet No. 21 (HMSO, 1964)

The Movement of Timbers. FPRL leaflet No. 47 (HMSO, 1965)

Strength and Bending Properties

The Strength Properties of Timbers. By G. M. Lavers. FPRL bulletin No. 50 (HMSO, 1967)

The Strength of Timber. FPRL leaflet No. 55 (HMSO, 1966)

The Steam-Bending Properties of Various Timbers. FPRL leaflet No. 45 (HMSO, 1958)

(Literature references continued)

Durability and Preservative Treatment

Decay of Timber and its Prevention. By K. St G. Cartwright and W. P. K. Findlay (FPRL, HMSO, 1958)

Insect and Marine Borer Damage to Timber and Woodwork. By J. D. Bletchly (FPRL, HMSO, 1967)

The Natural Durability of Timber. FPR record No. 30 (HMSO, 1959)

Non-pressure Methods of Applying Wood Preservatives. FPR record No. 31 (HMSO, 1961)

The Preservative Treatment of Timber by Brushing, Spraying and Immersion. FPRL leaflet No. 53 (HMSO, 1962)

Working and Finishing Properties

A Handbook of Woodcutting. By P. Harris (FPRL, HMSO, 1946)

Machining and Surface Finish. FPRL technical note No. 5 (HMSO, 1966)

Uses

Wood in Building for Purposes Other Than Structural Work and Carcassing. (Timber Research and Development Association, 1963)

The Design and Practice of Joinery. By J. Eastwick-Field and J. Stillman (Agricultural Press, London, 1961)

Wood Flooring (TRADA, 1959)

Wood Floors (TRADA, 1959)

Timbers for Flooring. FPR bulletin No. 40. (HMSO, 1957)

Hardwoods for Industrial Flooring. FPR Laboratory leaflet No. 48 (HMSO, 1954)

Timbers used in the Musical Instruments Industry (FPRL, 1956)

Timbers used in the Building and Repair of Railway Rolling Stock (FPRL, HMSO, 1956)

Timbers used in the Sports Goods Industry (FPRL, HMSO, 1957)

Timbers used in Cooperage and the Manufacture of Vats and Filter Presses (FPRL, HMSO, 1958)

Timbers used in Motor Vehicles (FPRL, HMSO, 1958)

Timbers used in the Boat Building Industry (FPRL, HMSO, 1964)

Timbers and Board Materials used in the Furniture Industry (FPRL, HMSO, 1966)

Index

This book is set in the Univers series and
printed in Great Britain by
The Journal Press (W. & H. Smith Ltd.)
Evesham, Worcestershire